An Evaluation of Japanese Environmental Regulations

Toshi H. Arimura • Kazuyuki Iwata

An Evaluation of Japanese Environmental Regulations

Quantitative Approaches from Environmental Economics

Springer

Toshi H. Arimura
Faculty of Political Science
and Economics
Waseda University
Shinjuku-ku, Tokyo, Japan

Kazuyuki Iwata
Faculty of Regional Policy
Takasaki City University of Economics
Kaminamie, Takasaki, Gunma, Japan

ISBN 978-94-017-9946-1 ISBN 978-94-017-9947-8 (eBook)
DOI 10.1007/978-94-017-9947-8

Library of Congress Control Number: 2015939693

Springer Dordrecht Heidelberg New York London

Printed on acid-free paper

Springer Science+Business Media B.V. Dordrecht is part of Springer Science+Business Media (www.springer.com)

Preface

Japan has overcome several severe environmental challenges. An obvious example is air pollution. Ambient air quality had deteriorated in the 1960s and 1970s, similar to the current situation in China and other Asian countries. It improved after regulations on emissions from stationary sources were implemented. However, in metropolitan areas such as Tokyo and Osaka, the problem persisted through the 1990s as a result of mobile sources. Japan overcame the problem of emissions from mobile sources by implementing a unique statute, the *NOx-PM Act*. This unique Japanese experience, however, has not been shared with the global community.

Another unique aspect of the Japanese environmental situation is the high level of commercial energy efficiency. The Japanese government introduced the *Energy Conservation Act* to promote energy efficiency. It is unclear, however, whether the observed efficiency gains are the result of private firms' self-motivation or of government energy efficiency policy.

Although numerous environmental regulations and policies have been implemented in Japan, few have been quantitatively evaluated. Therefore, it is unclear whether these environmental regulations have achieved their goals. In addition, even when evaluations were conducted, they lacked an economic perspective and a quantitative assessment. This lack of quantitative economic evaluation contrasts with the experiences of US and European countries, where quantitative economic analysis of environmental regulations is common. Although many expect Japan's environmental policies to improve the country's environmental quality, Japanese regulators have not examined whether these policies have improved environmental quality and social welfare or the extent to which each regulation contributed to any improvements. It is possible that the costs of certain environmental regulations have exceeded their benefits. Moreover, it is possible that the same environmental objectives could have been achieved with a more appropriate, and less costly, policy design. Therefore, it is important to examine the effects of Japanese environmental policies.

This book describes and analyzes several unique Japanese environmental regulations. Most chapters are based on an analysis that we conducted as committee members for national or local regulatory agencies. Our focus is on air pollution, energy efficiency, and climate change. To our knowledge, this is the first book to present evaluations of Japanese policies on these environmental issues from an economic perspective using a quantitative approach. This book's findings and suggestions for environmental policy evaluations will benefit not only policymakers and government officials in Japan but also those in developing countries and other developed countries, where the public faces similar environmental problems.

This book is written for undergraduate/graduate students and professionals who have an interest in environmental policy. It provides an introduction to Japan's environmental policies and regulations. Moreover, it offers economic analyses and Regulatory Impact Analyses (RIAs) of environmental regulations implemented or planned by the national government or local governments. Because we adopt a quantitative economic approach to these evaluations, readers may benefit from an introductory understanding of economics and statistics.

The first chapter of the book reviews the basics of environmental economics and provides the current status of RIAs in Japan. The following four chapters address automobile regulation with a focus on air pollution. Chapter 2 analyzes a national policy, known as the *NOx-PM Act,* which prohibits the use of old and polluting vehicles in metropolitan areas. Chapter 3 examines a regulation established by the Tokyo metropolitan government, which requires old trucks that fail to meet a specified emission standard to install pollution control equipment. Chapter 4 extends the scope of the previous two chapters. This chapter examines the impact of the *NOx-PM Act* in other areas, such as the used car market and used vehicle exports. Chapter 5 presents an economic analysis of a highway toll reduction and its environmental consequences. This research reveals that the toll reduction had an unexpected negative social impact because it increased traffic congestion and associated environmental problems.

The final three chapters address policies/regulations related to energy efficiency and climate change. Chapter 6 analyzes the effectiveness of Japan's *Energy Conservation Act*, which was originally introduced in 1979 and has been amended numerous times to address climate change. Specifically, this chapter empirically examines the effectiveness of the act in the hotel industry. Chapter 7 examines the impact of a proposed economy-wide carbon tax. The chapter examines the short-term economic impacts in each sector using an input-output analysis. This chapter also evaluates a reduced carbon tax for energy-intensive industries and examines its effectiveness. It also discusses the impact of the proposal on households. Finally, Chap. 8 discusses the role and limitations of economic models for evaluating Japan's mid-term GHG (greenhouse gas) emission target during the post-Kyoto period.

This book is based on the Japanese book, *An Evaluation of Japanese Environmental Regulation: Quantitative Approaches from Environmental Economics* published by Sophia University Press. It is not, however, a simple translation. We updated and revised the Japanese version to allow readers outside Japan to understand and appreciate its contents. Each chapter is partially or entirely based on a research paper published either in Japanese or English. The sources for each chapter are listed at the end of the book.

We are indebted to numerous individuals and organizations. The analyses in Chaps. 2 and 3 would not have been feasible without the automobile inspection and registration data provided by the Ministry of Internal Affairs and Communications. We are grateful to Ai Oba for assistance in collecting highway traffic data for the analysis in Chap. 5. We would like to thank the Agency for Natural Resources and Energy for its cooperation in supplying the data for the analysis in Chap. 6. The useful comments and information for Chap. 7 provided by the Tokyo Tax Commission are also appreciated.

We thank Professors Fukujyu Yamazaki and Yoshitusgu Kanemoto for their encouraging and helpful comments on many chapters of the book. We appreciate the assistance of Makoto Sugino and Minoru Morita with the analysis in Chap. 7, Hakaru Iguchi with Chap. 4 and Shiro Takeda for comments on Chap. 8.

Writing the English version required numerous revisions and updates of manuscripts. We are grateful to Hajime Katayama for helpful comments on several chapters of this book. We also have benefited from comments from Shigeru Matsumoto, Takaharu Eto, Shinya Horie, and Hanae Katayama.

Toshi Arimura would also like to thank Ed Foster for his guidance during his graduate studies. Without Foster's encouragement and lectures on public economics at the University of Minnesota, Arimura would not have had the opportunity to be engaged in policy evaluations for Japanese regulatory agencies or to be in a position to publish this book. Finally, Toshi Arimura dedicates this book to Joe and Betty Anderlik whose generosity as a host family supported his Ph.D. study in Minnesota. Without their hospitality, Arimura could not have survived Minnesota's cold winters.

Sources

The following chapters are partially based on the following sources:

Chapter 2: Iwata K, Arimura TH (2009) Economic analysis of Japanese air pollution regulation: an optimal retirement problem under vehicle type regulation in the NOx–particulate matter law. Transport Res Part D 14(3):157–167

Arimura TH, Iwata K (2008) Economic analysis on motor-vehicle type regulation: policy evaluation of NOx-PM law. Environ Sci 21(2):103–114 (in Japanese)

Chapter 3: Iwata K (2011) Cost-benefit analysis of enforcing installation of particulate matter elimination devices on diesel trucks in Japan. Environ Econ Policy Stud 13(1):1–19

Chapter 4: Iwata K, Fujii H, Managi S (2012) Does environmental regulation affect on outside of the regulated areas?: empirical analysis of Japanese automobile NOx-PM law. Rev Environ Econ Policy Stud 5(1):21–33 (in Japanese)

Chapter 6: Arimura TH, Iwata K (2008) The CO_2 emission reduction under the law concerning the rational use of energy: an empirical study of energy management in the Japanese hotel industry. Rev Environ Econ Policy Stud 1(1):79–89 (in Japanese)

Chapter 7: Sugino M, Arimura TH, Morita M (2012) The impact of a carbon tax on the industry and household: an input-output analysis. Environ Sci 25(2):126–133 (in Japanese)

Chapter 8: Arimura TH (2011) Evaluating Japanese carbon mitigation policies: the role of economic models. Energy Resources 32(2):9–13 (in Japanese)

Shinjuku-ku, Tokyo, Japan Toshi H. Arimura
Kaminamie, Takasaki, Japan Kazuyuki Iwata

Contents

1 Environmental Policy Evaluations in Japan: Concepts and Practice 1
1.1 Key Concepts in Environmental Economic Theory 1
1.2 Instruments for Environmental Policies 3
1.3 Evaluation of Environmental Policies 5
1.4 Approaches and Tools for Policy Management and Evaluation 8
1.5 Policy Evaluations in Japan in the Past and Present 13
1.6 Contents of This Book 14
References 17

2 Ex Ante Policy Evaluation of the Vehicle Type Regulation 19
2.1 Introduction 20
2.2 A Brief History of Japanese Air Pollution Regulation: Background of the VTR 21
2.3 Compliance Methods 24
2.4 Cost of the VTR 24
2.4.1 Compliance Cost 24
2.4.2 Opportunity Cost of Replacement 25
2.4.3 Income from Selling Old Vehicles 29
2.4.4 Compliance Cost 29
2.4.5 Total Cost 30
2.4.6 Potential Biases in *TC* 32
2.5 Benefit of the VTR 33
2.5.1 Effect of the VTR 33
2.5.2 Baseline: Emissions Without the VTR 34
2.5.3 Emissions with the VTR 34
2.5.4 Emissions Reduction 35
2.5.5 Net Benefit 37
2.6 Cost-Benefit Analysis 38

2.7 Efficiency of the VTR: Simulation of Alternative Policies 39
2.7.1 Marginal Abatement Cost 39
2.7.2 Simulation of Improved Efficiency 40
2.7.3 Discussion on the Optimal Simulation 45
2.8 Conclusion 46
Appendix 2.1 Average Life Expectancy: *L* (Year) 47
Appendix 2.2 The Number of Vehicles Replaced, NR^t 48
Appendix 2.3 Mileage Adjustment Rate 49
References 49

3 Cost-Benefit Analysis of Enforcing Installation of Particulate Matter Elimination Devices on Diesel Trucks 51
3.1 Introduction 51
3.2 Operational Regulation and Vehicle Type Regulation 53
3.3 Cost of Operational Regulation 57
3.3.1 Cost of Method 1 57
3.3.2 Cost of Method 2 58
3.3.3 Total Cost 59
3.4 Benefit of Operational Regulation 60
3.4.1 Emissions Without Operational Regulation 61
3.4.2 Emissions with Operational Regulation 62
3.4.3 Total Benefit of Operational Regulation 63
3.5 Cost-Benefit Analysis 63
3.6 Robustness Check 65
3.7 Conclusion 68
References 69

4 Does Environmental Regulation Affect on Outside of the Regulated Areas? Empirical Analysis of Japanese Automobile NOx-PM Act 71
4.1 Introduction 71
4.2 The Effect of VTR on Unregulated Areas 73
4.3 Estimation Models 76
4.4 Data and Explanatory Variables 78
4.5 Estimation Results 80
4.6 Conclusion 83
References 84

5 An Economic Welfare Analysis of the 1,000-Yen Expressway Discount 87
5.1 Introduction 87
5.2 Economic Welfare Analysis Using Average Costs 90
5.3 Generalized Equilibrium Demand for the Tomei Expressway and Social Cost 91
5.3.1 Changes Observed in the Tomei Expressway 91
5.3.2 Generalized Cost and Benefits for the Tomei Expressway due to the 1,000-Yen Expressway Discount 93
5.3.3 Average Social Cost and the Cost of the 1,000-Yen Expressway Discount for the Tomei Expressway 95

5.4 Changes in Costs . . . 97
5.4.1 Expressway Toll System . . . 97
5.4.2 Traffic Volume . . . 97
5.4.3 Traffic Congestion . . . 99
5.4.4 Time Cost Estimation . . . 100
5.4.5 Fuel Efficiency . . . 101
5.4.6 Fuel Consumption . . . 101
5.4.7 Other Mileage Costs . . . 102
5.4.8 External Cost: Environmental Cost . . . 103
5.5 The Cost and Benefits of the Tomei Expressway . . . 104
5.5.1 Generalized Cost . . . 104
5.5.2 Social Cost . . . 107
5.5.3 Cost and Benefits . . . 109
5.6 Aggregate Cost and Benefits of the 1,000-Yen Expressway Discount System . . . 111
5.6.1 Impact on the Tokaido Shinkansen . . . 111
5.6.2 Net Benefits from the New Expressway Discounts . . . 114
5.7 Conclusions . . . 115
Appendix 5.1 The Model for Toll Rates . . . 117
Appendix 5.2 Calculation of Traffic Volumes . . . 117
Appendix 5.3 The Model for Traffic Congestion . . . 118
Appendix 5.4 Time Cost Estimation . . . 119
Appendix 5.5 Fuel Cost Exclusive of Tax and Fuel Tax . . . 120
References . . . 121

6 The Evaluation of "Comprehensive Management Under the Act on the Rational Use of Energy" as a Measure to Combat Climate Change for the Hotel Industry . . . 123
6.1 Introduction . . . 123
6.2 The Act on the Rational Use of Energy (Energy Conservation Act) . . . 125
6.2.1 Overview . . . 125
6.2.2 Assessment Standards of the Energy Conservation Act . . . 128
6.2.3 Evaluation Target . . . 128
6.3 Estimation Methodology and Results . . . 129
6.3.1 Data and Evaluation Criteria . . . 129
6.3.2 Econometric Model and Data . . . 132
6.3.3 Results: The Effect of Being Raised to a Type 1 Facility . . . 134
6.3.4 Discussion . . . 136
6.4 Conclusions . . . 139
Appendix 6.1 Quantile Regression . . . 140
References . . . 143

7 Economic Impacts of the GHG Tax on the Japanese Economy: Short-Term Analysis 145
7.1 Introduction 145
7.2 The Energy-Related Taxation System in Japan and Key Points of the Proposed Carbon Tax 147
7.2.1 Energy-Related Taxation and the Discussion at the Tokyo Tax Commission 147
7.2.2 Key Issues in Carbon Taxation (the Global Warming Countermeasure Tax) 150
7.3 The Input-Output Model 151
7.3.1 Input-Output Analysis 151
7.3.2 Model 153
7.4 A Simulation Scenario and the Results of the IO Analysis 154
7.4.1 The Taxation Scenario 154
7.4.2 Results 156
7.4.3 The Effects of National and Local Taxes 158
7.4.4 The Effects of Carbon Taxation by Fuel Type 159
7.5 The Effects on Households 161
7.5.1 A Framework of the Impacts on the Household 161
7.5.2 The Impacts of the Carbon Tax on Household Expenditures with Different Levels of Income 162
7.5.3 The Impacts of the Carbon Tax on Household Expenditures in Different Regions 164
7.6 Conclusion 166
Appendix 7.1 166
References 173

8 The Impact on the Japanese Economy of Reducing Greenhouse Gas Emissions: The Role of Economic Models 175
8.1 Introduction 175
8.2 Economic Model Assessment of the Effects of GHG Emission Reductions on the Economy 176
8.2.1 The Benefits of GHG Emission Reductions 176
8.2.2 Economic Models for Evaluating the Economic Effects of GHG Emission Reduction Policies 177
8.2.3 Economic Effect Assessment: Policy Comparisons 178
8.2.4 Factors Affecting the Magnitude of Economic Impacts 179
8.2.5 Comparisons of the Models 179
8.2.6 The Importance of Effective Communication 180
8.3 Does Environmental Regulation Have Any Positive Effect on the Economy? 181
8.3.1 Possible Positive Effects on the Economy: Double Dividends 182
8.3.2 Possible Positive Effects of Regulation: The Porter Hypothesis 183

8.3.3 Japanese Product Sales in Overseas Markets 184
8.3.4 Green Consumption and Green Industry 184
8.4 Limitations of Economic Models 185
8.4.1 Assessment of Effects on the Household and Transportation Sectors 185
8.4.2 The Difficulty of Evaluating an Energy/Carbon Intensity Target 185
8.4.3 Assessment of Green Innovation 186
8.5 Conclusions 186
References 187

Index 189

Chapter 1
Environmental Policy Evaluations in Japan: Concepts and Practice

Abstract In this chapter, we will introduce key economic concepts and discuss methods for evaluating environmental policies. We will then briefly review policy evaluations in Japan. Finally, we will provide an overview of our quantitative policy evaluations presented in the rest of this book.

Keywords Welfare analysis • Environmental Regulation • Quantitative evaluation • Regulatory Impact Analysis • Policy evaluation in Japan

1.1 Key Concepts in Environmental Economic Theory

In economics, the total benefit for society is the sum of the benefit for consumers and that for producers.[1] A perfectly competitive market maximizes the total net benefit for the society at equilibrium. Hence, the market mechanism is considered to be efficient. This is not the case, however, when the consumption or production of goods comes with an *externality*. When production activity by a firm affects other firms or consumers directly, and not via market activities such as price changes, the activity is said to have an externality. Likewise, if the consumption of goods has effects on other consumers outside the market mechanism, the activity has an externality. Environmental problems such as climate change and air pollution are typical examples of *negative externalities*. Air pollution, for example, negatively influences individuals by causing health problems.

An externality leads to *market failure*. That is, it causes a competitive market to fail to achieve efficiency. If, for example, consumers use gasoline for driving automobiles, they emit carbon dioxide (CO_2) and cause greenhouse gas effects. Automobiles can also be a source of noise pollution. Although these negative externalities are the costs imposed on society, they may not be reflected in the prices of automobiles and gasoline. If so, the prices are lower than they should be and gasoline and automobiles are hence consumed more than at the optimal, socially desirable level. Externalities and market failure can also be observed in production activities. For example, production activities that consume fossil fuel

[1] This section discusses basic concepts in environmental economic theory that are useful in evaluating policies. Readers familiar with microeconomics may wish to skip to the next section.

T.H. Arimura, K. Iwata, *An Evaluation of Japanese Environmental Regulations*,
DOI 10.1007/978-94-017-9947-8_1

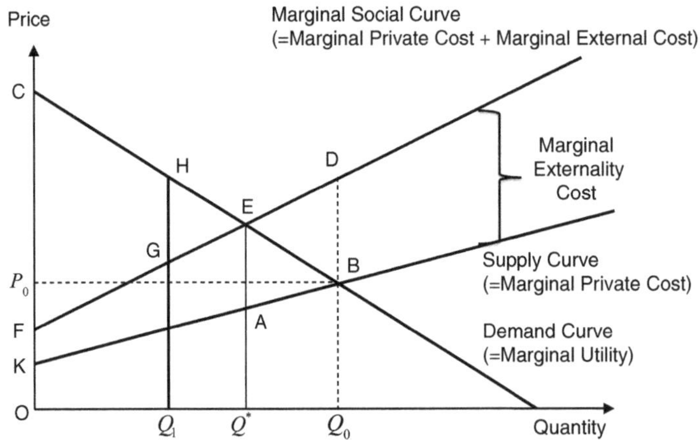

Fig. 1.1 Market with environmental problems

cause climate change by emitting CO_2 and also trigger air pollution by producing nitrogen oxides (NOx) or particulate matter (PM). If there is no governmental intervention in these activities, the quantity produced exceeds optimal levels and social welfare cannot be maximized.

These concepts can be illustrated with demand and supply curves as illustrated above in Fig. 1.1. In the figure, the downward sloping curve with the intercept C is the demand curve. The upward sloping curve with the intercept K is the supply curve. If there is no interference by any regulation, then, at the price P_0, the quantity demanded is equal to the quantity supplied, Q_0, in the market. Therefore, point E is the market equilibrium point, where P_0 and Q_0 are equilibrium price and equilibrium quantity, respectively.

Social welfare is measured by the sum of consumers' benefit and producers' benefit. Economics uses *consumer surplus* to capture consumers' benefit. Consumer surplus is the net benefit that consumers receive from consuming goods; in other words, it is the benefit that they receive from consuming goods minus the cost of purchasing them. The demand curve shows the marginal benefit that the consumer can enjoy from consuming each additional unit of a good. Consumer surplus is expressed in Fig. 1.1 as the area $\triangle P_0BC$, which is the area between the demand curve and the price line.

The supply curve shows the marginal cost that the producer pays for producing each additional unit of a good. The producer's benefit from the production activity is measured by *producer surplus*, which captures the profit (plus fixed cost). Note that the area below the supply curve is the production cost since the supply curve exhibits the marginal cost of the production. Moreover, the sales are expressed with the $\square\ OQ_0BP_0$. Therefore, producer surplus is depicted in the figure as the area $\triangle\ KBP_0$.

If there is no environmental problem associated with the consumption or production of a product, social welfare is $\triangle$ KBC, that is, the sum of consumer and producer surpluses. Here, social welfare is maximized as it achieves the equilibrium point B.

If the consumption or production of a good causes damage outside the market, such as environmental problems, it incurs external cost, i.e., a negative externality. Figure 1.1 shows the size of external cost by using the marginal external cost curve. Producers do not pay external costs; they only pay private cost, which is the cost for producing products. By adding producers' private cost and external cost, we can obtain social cost. Thus, we can obtain the marginal social cost curve by vertically adding the marginal private cost curve and marginal external cost curve. The marginal social cost curve passes through point F in Fig. 1.1. The area □ KBDF is the external cost caused by the equilibrium quantity Q_0.

Given this external cost, we need to subtract it from the sum of the consumer surplus and the producer surplus to calculate social welfare. Since some parts of the consumer surplus and the producer surplus are cancelled by the external cost, we can obtain social welfare as follows.

$$
\begin{aligned}
&\text{Social Welfare} = \text{Consumer Surplus} + \text{Producer Surplus} - \text{Externality Cost} \\
&= \triangle \mathrm{P_0BC} + \triangle \mathrm{KBC} + \square \mathrm{KBDF} \\
&= \triangle \mathrm{FEC} - \triangle \mathrm{EBD}
\end{aligned}
$$

If the government interferes with the market in a way that decreases the production level to level Q^*, social welfare becomes △ FEC, which is higher than that under the competitive equilibrium. Here, we can see that social welfare is increased.

Note that environmental issues (i.e., external costs) still exist even with government intervention. Now, what will happen if we decrease production to a level less than Q^* in order to decrease environmental damage? Suppose that we decrease the production to Q_1 in Fig. 1.1. Then social welfare becomes △FEC − △GEH = □FGHC, which is less than social welfare at Q^* by △ GEH. This implies that social welfare is maximized at Q^*.

Recalling our knowledge of microeconomics, we know that the demand curve represents the marginal benefit in a market. Therefore, we can find that social welfare is maximized when the following equation holds at Q^*:

$$\text{Marginal Benefit in a Market} = \text{Social Marginal Cost.}$$

1.2 Instruments for Environmental Policies

Ideally, environmental policies should be implemented in a way that maximizes social welfare. What kind of instruments can be utilized as environmental policies? This section will explain types of policy instruments and examine how they function by using demand and supply curves.

Policy instruments can be categorized roughly either as prescriptive regulations (also called as "command and control regulations") or economic incentives. The former has been used more commonly than the latter. The most traditional kind of prescriptive regulations explicitly stipulate specific technologies that must be used, such as scrubbers to remove SO_2. Another approach is for regulators to set performance standards that specify maximum pollution levels Environmental Protection Agency (EPA) 2014. Prescriptive regulations thus allow the government to directly place constraints on the activities of producers or consumers. In other words, prescriptive regulations do not provide incentives for producers or consumers. For example, as a result of specific technologies prescribed, producers do not have incentives to invent new technologies to reduce pollutions. Likewise, if a maximum pollution level is specified, incentives for producers to reduce pollution below the maximum level will be weak.

While not as widely used as prescriptive regulations, economic incentives such as emission fees and tradable permits are often advocated by economists. In theory, economic incentives are provided to profit-maximizing producers and utility-maximizing consumers who behave optimally in line with policy goals. If environmental costs are adequately reflected in the prices of goods and services, producers and consumers internalize adverse effects of their activities on the environment. That is, increases in production and consumption costs should result in a socially optimal level of production.

Figure 1.2 illustrates a market where an emission fee (hereafter referred to as an "environmental tax") is introduced. The market examined here is identical to that in Fig. 1.1 and so are the demand and supply curves. Environmental tax t is assigned per unit of production. t is equivalent to the size of the marginal externality, which corresponds to the segment AE in the figure. Producers must pay tax t in addition to their private costs for production. Thus, the marginal cost for producers increases, shifting the supply curve upwards to the curve that passes point L. The market equilibrium point then changes to point E, where the equilibrium price is P^* and the equilibrium quantity of consumption and production is Q^*.

The introduction of the appropriate tax rate t^* is expected to resolve, or at least alleviate, environmental problems by way of market mechanisms and economic

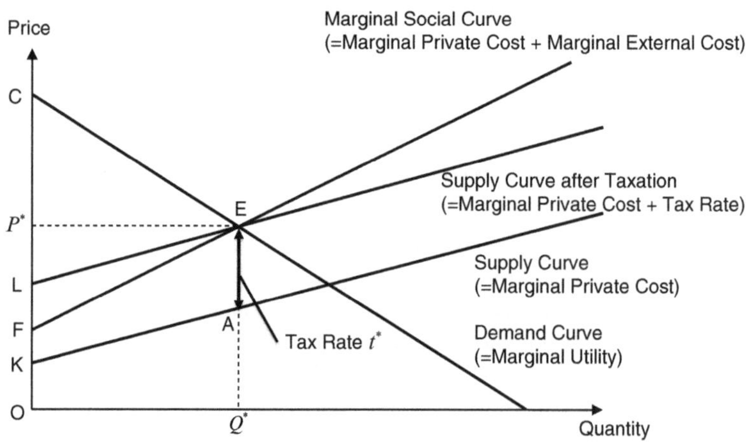

Fig. 1.2 Introduction of an environmental tax

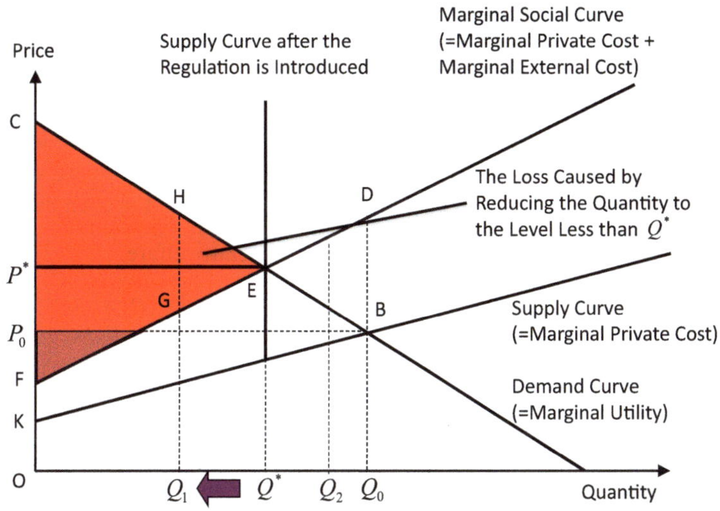

Fig. 1.3 Introduction of a regulatory instrument

agents' optimization. In practice, however, it is not easy to introduce economic incentives like environmental taxes. They are generally considered by producers and consumers as burdens to bear and, hence, they not easily approved. Even if an environmental tax is successfully introduced, the tax rate is usually set to be lower than the socially optimal level t^*.

Figure 1.3 shows the market where a regulative instrument is introduced. If the government limits the production to level Q^*, then social welfare can be maximized and becomes ΔFEC, that is, the same level as in the case where an environmental tax is introduced to the market. In reality, the regulation is not necessarily set to be Q^*. Consider a case where the production level is Q_0 under no regulation and the government then set a regulation to limit the production level to Q_2. The benefit created by the regulation exceeds the cost of the regulation, which means that the regulation increases social welfare and is therefore justified as a reasonable instrument. In fact, more stringent regulation is desired to further increase social welfare.

What happens if the regulation is enforced more stringently to the extent that the production is reduced to Q_1? When economic activities are excessively suppressed by a regulation, i.e., if the production level is suppressed to lower than Q^*, the cost exceeds the benefit, resulting in decreasing social welfare. Too much cost imposed on firms to reduce environmental problems is therefore not socially desired.

1.3 Evaluation of Environmental Policies

Building on the concepts presented above, we will now introduce essential inquiries to bear in mind when evaluating environmental policies. There are five inquiries, though they are not complete or exclusive: (1) whether a given environmental policy increases social welfare, (2) whether the policy is implemented efficiently, (3) to what extent the control subjects comply with the policy, (4) what, if any, are the positive and negative byproducts of the policy, and finally, (5) whether the burden of the policy is evenly distributed among the control subjects. These inquiries will be explored in the remaining chapters of this book where we conduct quantitative evaluations of selected environmental policies. Below, we will briefly overview the inquiries with reference to the studies presented in each chapter.

1. Does the given environmental policy increase social welfare?

One of the objectives of evaluating a policy of any kind is to determine whether benefits of the given policy exceed its costs. Obviously, it is pointless to implement a policy that does not increase, but decreases, social welfare. The same applies to environmental policies. The benefits and costs of an environmental policy must be quantified to find out its validity.

As we discussed with Fig. 1.3 in the previous section, we can examine the effect of a regulatory policy by seeing how much production and consumption increase or decrease. If the current regulation sets the quantity at Q_2, it should be strengthened to the optimal level Q^* so that the social surplus can be maximized. If it is at Q_1, then the regulation should be relaxed to Q^* so that the quantities of production and consumption can be increased. In this way, quantitative analysis allows us to see whether a policy is implemented in a way that maximizes social surplus. Up to the present, only a small number of academic or government-led studies have evaluated environmental policies in Japan from a quantitative perspective. The studies presented in this book provide examples of quantitative analysis on environmental regulations in Japan, including the Automobile NOx/PM Act (Chaps. 2, 3, and 4).

2. Is the policy implemented efficiently?

In Sect. 1.2, we implicitly assumed that marginal abatement costs are equalized among polluters and that the target of an environmental policy is achieved efficiently. Economic incentives such as an environmental tax or tradable permits equalize marginal abatement costs among polluters through price mechanisms in the market (Kolstad 2010). In this case, pollution is reduced most efficiently, that is, with the least social cost. This is called the *equimarginal principle*. This efficiency is difficult to achieve with prescriptive regulations because it is hard for the government to acquire information on marginal abatement costs among polluters.[2] Even if pollution is reduced by prescriptive regulations, we are not sure whether it can be achieved with the least cost. This is why we need to examine whether policies achieve their goals efficiently. The issue of efficiency will be discussed in Chaps. 2 and 3.

[2] See Chapter 4 in Kolstad (2010).

3. Policy target and compliance

Producers and consumers may not necessarily comply with environmental regulations, especially if their economic activities are not closely monitored by the government. The incentive to comply with a regulation may also be weak if penalties for incompliance are not stringent. This is exemplified by drivers disobeying speed limits and the illegal dumping of industrial waste. Chapter 6 examines the extent to which the hotel industry in Japan complies with the energy efficiency target under the Energy Saving Act.[3]

4. Ancillary Benefits and Costs

When evaluating an environmental regulation, one must consider whether the regulation generates *ancillary benefits*, that is, byproducts or benefits other than its originally intended effects (Harrington et al. 2009). Suppose, for example, that a carbon tax is introduced to reduce GHG emissions. Given gasoline prices that include such a tax, utility-maximizing consumers will reduce gasoline consumption and thus NOx and PM emissions will also be reduced. This, in turn, decreases GHG emissions and reduces air pollution. Here, the ancillary benefit of the carbon tax is the reduction of air pollution.

Ignoring ancillary benefits leads to underestimating a regulation's benefit for society. This weakens the regulation's validity and may fail to justify its implementation when it is desirable. Sometimes, ancillary benefits constitute a significant portion of the total benefit. Burataw et al. (2003) showed, for example, the ancillary benefit of a $25 carbon tax in the U.S. can be from $12 to $14. That is, the ancillary benefit covers the half of the cost. Chapter 4 will examine ancillary benefits of the Automobile NOx/PM Act.

Sometimes a policy or regulation may have ancillary costs. There has been little attempt to quantitatively evaluate policies that are likely to have generated ancillary costs. We will present in Chap. 5 a case study on a policy informally known as the "1,000 yen Highway Policy" and examine the ancillary costs of this policy.

5. Distributional Consequences: Burden and Equity

In the long run, environmental policies provide benefits by mitigating environmental problems and reducing external costs. In the short run, however, regulations may cause some burden for consumers and producers. For example, it is not hard to imagine that consumers will face higher prices of products if an environmental tax is introduced. Likewise, producers are likely to pay higher costs to purchase the same amount of inputs than before the tax is introduced. Therefore, it is necessary for the regulator to account for the burden of a regulation before its introduction.

Also, even if a policy is expected to be efficient in terms of social welfare, it may need to be revised in terms of equity if only particular industries or consumers have to bear the burden caused by the policy. In such a case, it is hard to reach an agreement among different social groups to introduce the policy. At the end, the

---

[3] Its official name is the Act on the Rational Use of Energy.

policy may fail to be implemented. Distributional issues should be taken into consideration when we design a policy in order to avoid a biased distribution of burdens among social groups.

As a way to evaluate a proposal for an environmental tax, in Chap. 7, we will conduct an *economic impact analysis* (EIA) and examine the burden of the tax for industries and households. We will address the issue of distribution by examining revisions made to the tax proposal.

1.4 Approaches and Tools for Policy Management and Evaluation

When planning an environmental policy, it is not easy to predict if the policy will indeed achieve its target as planned. Even after the policy has been implemented and its target has been achieved, it is uncertain if the target was achieved efficiently and, also, if the strength of the regulation is just right or should be adjusted. It is important to conduct policy evaluations on a regular basis in order to obtain up-to-date information on how the regulations are currently operating and what can be done to manage or improve the quality of the regulations.

Policy evaluations by the Japanese government incorporate *the PDCA cycle* to monitor the quality and effectiveness of a policy *before* its introduction ("ex-ante") and *after* it has been carried out ("ex-post").[4] The PDCA cycle constitutes four steps, as signified by each letter of the name. The first step is planning a policy (Plan), the second is to actually conduct the policy (Do), the third is to check the outcome of the policy (Check), and finally, to reconsider the policy (Action). Figure 1.4 below illustrates how a policy is evaluated in accordance with the PDCA cycle. In the cycle, a policy is examined by two types of assessment: ex-ante and ex-post evaluations. The former is conducted at the first step (Plan), and the latter at the final step (Action).

An ex-ante evaluation, or a policy evaluation at the planning step, helps us foresee how the introduction of a policy influences its society and whether it will improve social welfare. In contrast, an ex-post evaluation, i.e., policy evaluation at the step of reconsidering, helps us identify the actual effects caused by the introduction of a policy. The effects of a policy may not be sustained at the same levels over time because economic situations often change. Estimation results may also change over time if parameters used in the evaluations change. For example, if the condition of the economy changes from a recession to a boom, we may use different discount rates to evaluate identical policies. Therefore, it is ideal for every policy to be re-evaluated when some time has elapsed after its implementation. If we find that

[4] In this book, "ex-ante" means time before the policy is introduced, and "ex-post" means time after the policy is introduced. Note that "ex-post" does not mean "after the policy is completed or ended." An ex-ante/ex-post policy evaluation is also called a "regulatory impact analysis (RIA)" when the object of the evaluation is a regulation.

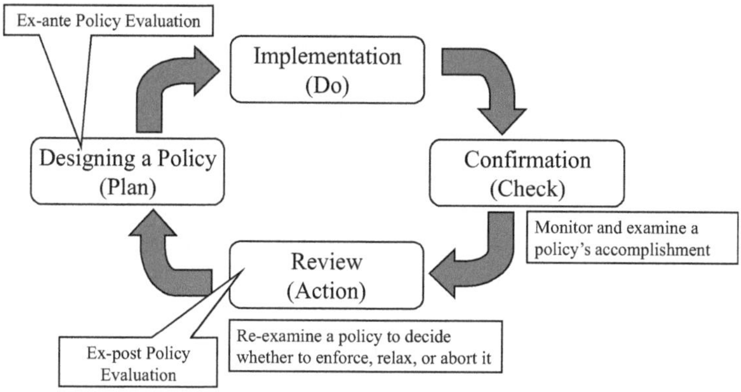

Fig. 1.4 PDCA cycle

a policy cannot deliver desired results, we may determine to relax, suspend, or permanently abandon it. At the same time, as part of an ex-post evaluation, we should diagnose the problems that prevent the policy from achieving expected results.

If an ex-post evaluation indicates that the policy successfully conveys results that are better than expected, then it will be continuously used, or in some cases, it might be enhanced. The evaluation is supposed to include prescriptions showing which parts of the policy should be changed to make it even more efficient. In this way, ex-post evaluations play a critical role in revising the original policy or developing an alternative policy.

Next, we turn to the methodological issue of how policies are evaluated.[5] We can categorize approaches for evaluating policies into four types: (1) qualitative evaluation, (2) quantitative evaluation, (3) cost effectiveness analysis, and (4) cost benefit analysis. For the purpose of understanding what we can do with each methodology, we will consider the following scenario. Suppose that plants emit sulfur dioxide (SO_2) and the government wants to control this plant pollution. Two

[5] We need to consider what we should use to measure the effects of a policy. For example, emission standards for automobiles include limits for detrimental substances per kilometer traveled. Should we measure the regulation's accomplishment by the degree of the compliance by automakers, i.e., the regulation's direct effect? Or, should we measure by the improvement of the ambient air quality caused by the regulation, which is the regulation's indirect effect? We call the first measurement "output" and the second "outcome." Policymakers usually formulate a policy aiming to obtain effects more than the policy's direct effects, i.e., effects that result from the control subjects complying with the policy. In other words, indirect effects are expected at the time a given policy is being prepared. Thus, it is appropriate to measure the effects by outcome. In this book, we will measure policies' effects by their outcomes. For more details on the measurement for effects, see Hatry (2007).

regulatory approaches are available and the government wants to know which one should be adopted: a *business suspension order* or *business improvement order*.[6]

Qualitative policy evaluations consider possible outcomes of implementing each *order* from various perspectives. For example, by conducting interviews both with plant employers, plant employees, and residents in the plant area, an evaluation report makes their voices and concerns heard. Information collected by this type of research is useful for pointing out the negative outcomes of shutting down plants, such as layoffs and the loss of profit from reducing production, as well as positive outcomes associated with reducing pollution.

During Japan's high-growth period (from the 1950s to the early 1970s), business suspension orders were issued to plants partially based on qualitatively examining pollution from the plants. Presumably, the orders increased social welfare by shutting down plants and reducing external costs caused by the plants. However, qualitative analysis does not provide any supporting evidence that the benefit of closing down plants exceeded the cost of doing so. In addition, the analysis does not use an objective measure that enables us to compare between outcomes of the suspension order and those of the improvement order. Hence, we cannot easily determine which order is more reasonable based only on the evaluation results.

In contrast, the other three methodologies evaluate the two orders quantitatively and thereby provide objective measures for us to compare them.[7] The simplest quantitative evaluation examines effects of a policy without considering its associated costs. For convenience, we temporarily call this approach "effects analysis." This method is frequently utilized in many policy evaluations in Japan because it is less data demanding compared to the other quantitative methodologies. As shown in Fig. 1.5, effects analysis identifies the amount of SO_2 reduction that can be

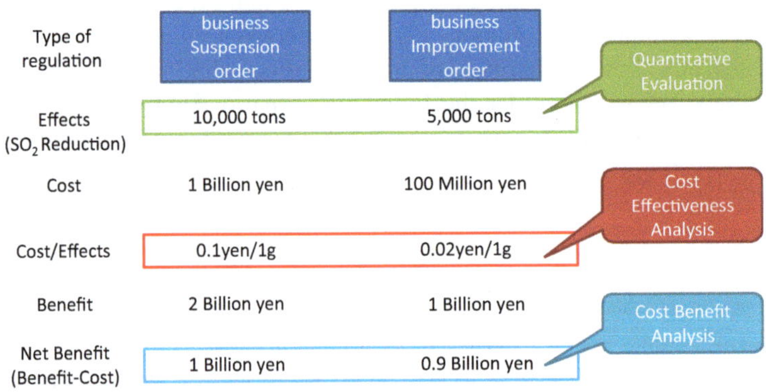

Fig. 1.5 Quantitative evaluations

[6] It is a type of administrative punishment that orders the pollution source to take action to reduce pollution caused by its production activities.

[7] Boardman et al. (2010) is an excellent reference for the economic methods to evaluate policies in general.

achieved by the suspension order and the improvement order, respectively. Let us assume that the predicted reduction of SO_2 is 10,000 tons by the suspension order and 5,000 tons by the improvement order. Based on this result, one may conclude that the government should choose a suspension order over an improvement order. What we did with the analysis is merely compare the amount of SO_2 reductions possibly achieved by each order. The analysis does not provide any information on the costs incurred by the respective orders and, hence, we cannot tell whether the improvement order is in fact a sensible policy decision.

Instead of focusing only on the positive effects associated with each order, we want to know both the positive and negative effects. It is possible to do so with the approach called *cost effectiveness analysis*. Social costs associated with a regulation are often not clear. However, one can estimate social costs if relevant data are available. In the above scenario, suppose that the following information is available about the social costs of the respective orders: the suspension order and the improvement order incur social costs by 1 billion yen and 100 million yen, respectively. The effects of each order remains the same, i.e., 10,000 tons of SO_2 can be reduced by the suspension order and 5,000 tons by the improvement order. Based on these numbers, we now know that the suspension order can reduce 10 g of SO_2 by incurring a unity of yen, whereas the improvement order reduces 50 g of SO_2 per yen. Thus, we can conclude that the improvement order is more desirable for this society.

It should be noted that cost effectiveness analysis has limitations in that it compares policies only with respect to the amount of SO_2 reduction per unit of social costs. It gives no information about whether the benefit from a regulation exceeds the costs of the regulation. We can see from the analysis that the improvement order wins over the suspension order in terms of cost effectiveness. That does not necessarily mean, however, that the improvement order is worthy of being implemented in practice.

Cost benefit analysis, an improved version of cost effectiveness analysis, enables us to compare the benefits and costs associated with a policy.[8] Cost benefit analysis converts the positive and negative effects of a policy into monetary terms, which we refer to as the benefits and costs of the policy, respectively. In this scenario, let us further suppose that the external costs caused by SO_2 emissions are 200,000 yen per ton of SO_2. The effects of the suspension order, 10,000 tons of SO_2 reduction, can also be converted to two billion yen of benefits. Likewise, the effects of the improvement order, 5,000 tons of SO_2 reduction, can be converted to one billion yen of benefits.

When we purely compare the amount of *net benefits* (i.e., the benefits minus the costs) brought by each order, the suspension order wins over the improvement order: one billion as opposed to 0.9 billion yen. Yet, if the policy maker uses the b/c criterion (the benefit-cost ratio) to choose between the two orders, the improvement

[8] See Boardman et al. (2010) and EPA (2014) for basic concepts of cost benefit analysis.

Table 1.1 Evaluation methods

Evaluation Method	Require information about cost	Compare among alternatives in terms of efficiency	Examine policy's validity (= whether social welfare increases)
Qualitative analysis	No	No	No
Quantitative analysis	No	No	No
Cost effectiveness analysis	Yes	Yes	No
Cost benefit analysis	Yes	Yes	Yes

order is better. While the suspension order is estimated to yield two billion yen of benefit, this can be achieved at the cost of one billion yen. Hence, the benefit-cost ratio is 2. The benefit of the improvement order, 1 billion yen, can be achieved at the cost of a hundred million yen and, thus, the benefit-cost ratio is 10. The point here is that the analysis does not tell the policy maker which policy is better but can provide relevant information that helps the policy maker make a decision.

Table 1.1 summarizes the four evaluation methods discussed above. Cost benefit analysis is most informative and should be applied to as many policy evaluations as possible. However, the analysis requires much more information than the other methods. In order to conduct a cost benefit analysis in the above scenario, for example, we need to gather data on the effects of the given policy, social costs of the policy, and the external costs of the pollution that the policy seeks to reduce. The more information required for evaluation, the more complicated the process of the evaluation becomes, which should result in more evaluation costs and time.

Problems associated with cost benefit analysis such as costs, time and the complicated process of estimation can be reduced by giving guidelines beforehand. The Japanese Ministry of Land, Infrastructure, Transport and Tourism (MLIT 2004) provides such guidelines to evaluate public sector infrastructure projects, including the value of the discount rate to be used and the definition of an infrastructure's benefit. These kinds of guidelines are relatively easy to apply to public sector projects whose costs are often made explicit, such as expenses of construction and maintenance. It is not so straightforward with regulations and policy instruments, however, because their costs depend on how economic agents, i.e., the objects of a regulation, respond to the regulation. Guidelines for cost benefit analysis are issued by the Japanese Ministry of Internal Affairs and Communication (2005, 2007a), though compared to those for public sector projects (MLIT 2004), they need to be refined, especially with regard to the definition of costs and the calculation method.

Conducting cost benefit analysis is particularly challenging when the scope of analysis is environmental regulations like the ones on pollution. It is difficult to measure external costs associated with pollution for the following reasons. The causal relationship between pollution and damage caused by pollution is not scientifically proven in many cases. Even if the scientific relationship is clear, valuation of the environment is difficult, since there are no markets for the environment. These uncertainties and challenges involved in cost benefit analysis reduce the credibility of the estimation results. To compensate for this problem, one can examine how sensitive the results are to the changes in parameters that represent external costs. This "sensitivity analysis" enhances the credibility of estimation.

1.5 Policy Evaluations in Japan in the Past and Present

Policy evaluations from an economic perspective hardly exist in Japan except for discussion of the GHG emission reduction target. This is because policy makers have not paid much attention to a policy's costs and its cost effectiveness. Rather, they have been interested in the effects of a policy, e.g., the reduction of damage caused by pollution.[9] The introduction of policy assessment is thus a new endeavor in Japan. This is evident especially when it is compared to cases in the U.S. where policy evaluations have relatively long tradition. In the U.S., policy assessment was initiated in 1968 with a scheme called the "Planning, Programming, and Budgeting System (PPBS)" Haveman, 1987). Under the Government Performance and Result Act (GPRA) in 1993, it was mandated that all federal governmental agencies pursue policy evaluations and disclose the results.

In 1998, thirty years after the introduction of assessment in the U.S., policy evaluations were initiated in Japan as part of the Government Reorganization Act.[10] The Government Policy Evaluations Act (GPEA) was then implemented in 2002. GPEA established a legal basis for policy evaluations, requiring both ex-ante and ex-post evaluations be conducted on a policy if it satisfies particular standards. This prevented each government office and ministry from evaluating their policies according to their own standards, which used to be the convention for policy assessment. GPEA also requires that evaluations be conducted by the Administrative Evaluation Bureau (AEB) of the Ministry of Internal Affairs and Communications (MIC) if a policy is implemented across offices and ministries.

---

[9] Among various policy instruments such as tax, subsidy and standards, prescriptive regulations perform best in securing the reduction of environmental damage. In Japan, prescriptive regulations were frequently implemented to control environmental pollutions during the 1970s (e.g., smoke control under the Air Pollution Control Law). Prescriptive regulations are used even today when urban pollutions must be urgently resolved. Although economic incentives are more efficient than prescriptive regulations, they are rarely introduced. Advantages and disadvantages of these instruments are discussed in textbooks of environmental economics (e.g., Kolstad 2010), along with comparisons among differing policies in terms of their efficiencies.

[10] See the Institute of Administrative Management (2006) for further information on the history of policy evaluations in Japan.

Article 9 of GPEA states that ex-ante evaluations are required for the following policy areas: research and development, public projects, official development assistance (ODA), and regulations.[11] Although it is desirable that all policies are subject to evaluation, GPEA designates exempt policy areas as follows: tax, subsidy, insurance annuity, judiciary proceedings, and the Self-Defense Forces Act.[12]

Given the legal basis for evaluation, what percentage of policies has actually been evaluated and in what ways have assessments been conducted? It is found that qualitative analysis is by far the most common approach to evaluating regulations in Japan. In three years, from October 1st of 2004 to September 30th of 2007, 247 ex-ante/ex-post evaluations were conducted on regulations (MIC 2007b). Only 26 of them (10.5 %) were quantitative analyses, as opposed to the majority (89.5 %, 221 evaluations) being qualitative. In addition, although regulations' effects are discussed in all evaluations, social costs are mentioned qualitatively in 184 evaluations (74.5 %), quantitatively measured in 23 evaluations (9.3 %), not mentioned in 8 evaluations (3.2 %), with 32 evaluations (13 %) not corresponding to any of the three. Note also that 232 out of 247 evaluations (93.9 %) conduct comparisons between the given regulation and other alternatives, though such comparisons cannot be conducted easily based on the results of qualitative evaluation (MIC 2007b).

Qualitative analysis aims to provide in-depth understandings of a particular issue and, thus, findings obtained from the analysis are not necessarily generalizable for other cases. The fact that the majority of regulations have been evaluated only qualitatively indicates that the research outcomes lack in reliability and validity. To get a more comprehensive picture of regulations that are currently in operation, it is necessary that evaluation incorporates perspectives besides qualitative assessment. Quantitative analysis that objectively estimates effects (benefits) and costs of a regulation must be introduced urgently in order to enhance the quality and credibility of policy evaluations in Japan.

1.6 Contents of This Book

In the rest of this book, we will quantitatively evaluate selected environmental regulations or policies. Below we will summarize the content of each chapter and conclude our introduction to the book.

---

[11] Ex-post evaluations are required by Article 8 of GPEA as follows: "An administrative organ shall carry out the ex-post evaluation based on the basic plan and the operational plan (http://www.soumu.go.jp/main_content/000082216.pdf)."

[12] The exempt policy areas are added as part of the revisions made to the Administrative Reform Act Ordinance, which was issued by MIC on October 1, 2007.

Chapter 2 will examine the Automobile NOx/PM Act, a regulation implemented in 2001 to control air pollution in Tokyo, Osaka, and Nagoya. The regulation is unique in that it enforces early retirement of older vehicles and mandates the use of newer vehicles in the three metropolitan areas. In the chapter, we will construct an economic model and conduct an ex-ante evaluation of the regulation. It is found that the regulation increased social net benefit. We will point out, however, that even a greater amount of benefit could be obtained if the policy makers schedule the time for retiring vehicles more effectively by incorporating results from the ex-ante evaluation. The importance of ex-ante impact analysis will be discussed along with the findings.

Chapter 3 will continue to examine the Automobile NOx/PM Act but focus on the requirement to install on old trucks the emission elimination devices called diesel particulate filters (DPFs). Local governments in Tokyo introduced an air pollution regulation requiring that old truck owners install DPFs on their trucks. We will construct an ex-ante evaluation model by extending the one used in Chap. 2 and conduct a cost benefit analysis. It will be argued that the cost of the regulation is lower than its benefit. Nonetheless, a further analysis will show that policymakers should opt for promoting clean vehicles via subsidies rather than by distributing DPFs.

The Automobile NOx/PM Act has ancillary effects on secondary vehicle markets. It is possible that vehicles forced to retire under the Act might have been transported and sold in secondary vehicle markets outside the regulated areas. It is also possible that the vehicles are exported overseas, particularly to developing countries and newly industrialized countries where emissions regulations are not as strict as in Japan. These possibilities will be explored in Chap. 4. We will investigate whether the regulation encouraged the outflow of old vehicles and then influenced secondary vehicle prices in unregulated areas. We do this by comparing secondary vehicle prices before and after the regulation's implementation. It is found that the difference in the prices is not statistically significant, indicating that the outflow of retired vehicles from the regulated to unregulated areas was not considerably large. Meanwhile, the export volume of secondary vehicles is found to have increased greatly after the implementation, suggesting the possibility that banned vehicles have been exported overseas. This exemplifies the pollution haven hypothesis. That is, the regulation encourages exporting old vehicles and thereby slows down the increase of new low-emission vehicles in the importing countries. It will be argued that policy makers need to consider the potential effect of an environmental regulation on areas that are not directly regulated. This is especially true if a regulation controls the use of particular goods, like vehicles, whose secondary markets are well established.

In Chap. 5, we will shift our attention from the Automobile NOx/PM Act and examine the highway toll reduction policy with regard to its economic and environmental impacts. In 2009, highway tolls in Japan were significantly reduced as part of an economic stimulus package for the Japanese economy. The price reduction led to a dramatic increase of traffic on highways across the country, incurring a negative externality of air pollution. In the chapter, we will assess the policy from an economic perspective by employing a general equilibrium model. We will discuss the policy's negative impact in terms of social welfare: the policy increased CO_2 emissions, air pollution, and traffic congestion that resulted in lost time for highway users.

The remaining chapters (Chaps. 6, 7, and 8) will explore policies on energy consumption and climate change. Chapter 6 will analyze the effects of the energy-saving act on the hotel industry. In response to the oil crisis of the 1970s, the Japanese government introduced the energy-saving act in 1979 to promote the rational use of energy. The act was originally meant to cope with energy security issues. However, in order to reduce greenhouse gases (GHG) and achieve the aims of the Kyoto Protocol, the government amended the act in the 1990s and required the hotel industry to report their energy consumption. The act also specifies a target of a one-percent annual reduction with the aim to decrease the energy intensity of fossil fuels and electricity. It is not clear, however, whether the targets have been met and reductions have been achieved. This is a question worthy of investigation, especially given that there are no severe penalties for not complying with the act. In the chapter, we will first provide an overview of the act and then examine its effects by using energy consumption data from the hotel industry. Our analysis will reveal that the act contributed to GHG emissions reduction. It is also found that while some hotels did not achieve the target, others reduced their consumption more than required. The results imply that the act has some room for improvement.

In Chap. 7, we will examine the economic impact of a carbon tax. A carbon tax is a market-based countermeasure for mitigating GHG emissions and, theoretically, it reduces GHG emissions at a minimum cost. Whether the tax should be introduced has been a topic of debate and discussion in many countries, including Japan. There are objections mainly because the introduction of the tax can largely increase the burden of costs for particular industries, such as the coal and petroleum industries, that use energy extensively. In the chapter, we will consider a proposal made at the Tokyo Tax Commission to introduce a carbon tax as a local tax. Specifically, we will examine the economic impacts of the tax on individual industries by using an input-output framework. We will discuss ways in which energy-intensive sectors can avoid or mitigate their cost burdens.

Chapter 8 will discuss how policymakers can utilize evaluation results, specifically in the context of developing GHG mitigation policies. In Japan, a series of debates was held with regard to post-Kyoto negotiations on climate change policies. Upon facing the end of the first commitment period of the Kyoto protocol (2008–2012), the Japanese government initiated discussion regarding mid-term GHG emission target levels after the second commitment period (2013–2020). For policymakers, it is particularly important to predict what effects the emission target might have on the Japanese economy. Various economic models have been used to analyze the impacts of differing emission targets. Different predictions were provided, creating confusion in the policy making process. The last chapter will consider how one can interpret and use results based on economic models when developing environmental policies.

References

Boardman A, Greenberg D, Vining A, Weimer D (2010) Cost-benefit analysis, 4th edn, Pearson series in economics. Prentice Hall, Upper Saddle River

Burataw D, Krupnick A, Palmer K, Paul A, Toman M, Bloyd C (2003) Ancillary benefits of reduced air pollution in the US from moderate greenhouse gas mitigation policies in the electricity sector. J Environ Econ Manage 45(3):650–673

Environmental Protection Agency (2014) Guidelines for preparing economic analyses. United States Environmental Protection Agency. http://yosemite.epa.gov/ee/epa/eerm.nsf/vwAN/EE-0568-50.pdf/$file/EE-0568-50.pdf. Accessed 15 May 2015

Harrington W, Heinzerling L, Morgenstern R (2009) Reforming regulatory impact analysis. Resources for the Future. Washington, DC

Hatry HP (2007) Performance measurement: getting results. Urban Inst Press, Washington, DC

Haveman RH (1987) Policy analysis and evaluation research after twenty years. Policy Stud J 16(2):191–218

Institute of Administrative Management (2006) Handbook of policy evaluation: advent of new era of evaluation. Gyosei, Tokyo (in Japanese)

Kolstad CD (2010) Environmental economics, 2nd edn. Oxford University Press, New York

Ministry of Land, Infrastructure, Transport and Tourism (2004) Technical guideline for cost benefit analysis of public project evaluation (in Japanese). http://www.mlit.go.jp/kisha/kisha04/13/130206/04.pdf. Accessed 15 May 2015

Ministry of Internal Affairs and Communications (2005) Guideline for implementation of policy evaluation (in Japanese). http://www.soumu.go.jp/main_content/000152600.pdf. Accessed 15 May 2015

Ministry of Internal Affairs and Communications (2007a) Guideline for implementation of ex-ante evaluation of regulations (in Japanese). http://www.soumu.go.jp/main_sosiki/hyouka/seisaku_n/pdf/070824_2.pdf. Accessed 15 May 2015

Ministry of Internal Affairs and Communications (2007b) Status of trial implementation of regulatory impact analysis (RIA) (in Japanese). http://www.soumu.go.jp/main_content/000076915.pdf. Accessed 15 May 2015

Chapter 2
Ex Ante Policy Evaluation of the Vehicle Type Regulation

Abstract In this chapter, we conduct an ex ante quantitative policy evaluation of the vehicle type regulation that aims to control vehicle emissions. Imposed under the Automobile Nitrogen Oxides–Particulate Matter Act, the regulation promotes the earlier replacement of old, high-emissions vehicles by prohibiting the use and registration of old vehicles in metropolitan areas. Using vehicle registration data, we identify all vehicles subject to the regulation and estimate the cost of the regulation, namely, the opportunity cost of earlier replacement. The total cost is estimated to be 521 billion yen. In addition, the regulation clearly delivers benefits, including the health benefit resulting from reduced emissions. We calculate the extent to which emissions from all regulated vehicles have decreased. Specifically, using the estimates of the marginal external costs of air pollutants, we measure the benefit of the regulation, which is found to be 1,202 billion yen. Therefore, the regulation is a reasonable policy whose net social benefit exceeds its costs; the net benefit (i.e., benefits less costs) is 681 billion yen. However, the results also show that the marginal abatement cost differs substantially across polluters, suggesting that the cost is not minimized. To examine the extent to which social surplus could be increased, we conduct a simulation in which we change the terminal year for each vehicle type. The social surplus is found to double in our simulation results, indicating that the regulation could have been conducted far more efficiently if its costs had been thoroughly examined prior to implementation. Our results also suggest that the regulation is currently less effective than economic instruments such as emissions taxes, as optimal and efficient environmental policy can be realized by introducing an environmental tax.

Keywords Automobile NOx-PM Act • Air pollution • Cost-benefit Analysis • Optimal Regulation • Ex-ante Evaluation

T.H. Arimura, K. Iwata, *An Evaluation of Japanese Environmental Regulations*,
DOI 10.1007/978-94-017-9947-8_2

2.1 Introduction

In this chapter, we conduct an ex ante policy evaluation of the vehicle type regulation (VTR) that was introduced under the NO_x -PM Act.[1] Based on the information available before the regulation was imposed, we estimate the cost and benefit of the VTR and then compare them. When considering the introduction of a prescriptive regulation, the regulator typically considers its benefit only (see Chap. 1). In contrast, we consider both the benefit and cost of prescriptive regulations (in this case, the VTR).

The regulator could have conducted the cost-benefit analysis in this chapter at the time of implementation because all of the information used in this study was already available at that time. Nonetheless, the regulator estimated the emissions reduction potential of the VTR without accounting for the implementation cost. By performing an ex ante evaluation of the VTR, we demonstrate the extent to which the social surplus produced by the VTR may have increased if its costs and benefits had been thoroughly examined and compared prior to implementation.

The VTR mandates the earlier replacement of old, high-emissions vehicles with new vehicles by designating retirement timing or "terminal years." Earlier replacement of old vehicles is expected to improve air quality because nitrogen oxide (NO_x) and particulate matter (PM) emissions intensities are lower for new vehicles than for old vehicles.

Several studies have examined vehicle emissions policies that promote the replacement of old vehicles. Lumbreras et al. (2008) showed, through simulation, that the earlier replacement of old vehicles is an effective policy measure to control air pollution. Other studies, such as Dill (2004) and Alberini et al. (1995, 1996), examined voluntary retirement programs in the U.S. and reported their effectiveness in reducing air pollutants. The present chapter focuses on the VTR, which, unlike other policy instruments, directly mandates when old vehicles must be retired.

The approach that we adopt in this chapter and in subsequent chapters differs from that in Sect. 1.1 of Chap. 1, in which we discussed the maximization of total social surplus. Specifically, in each chapter, we conduct a cost-benefit analysis that focuses exclusively on (1) the cost of complying with the VTR and (2) the benefit from the VTR measured in terms of the reduction of a negative externality (i.e., the health effects of air pollution). In other words, the effectiveness of the VTR is evaluated on the basis of its net social benefit, which is defined here as health benefits less compliance costs.

We employ this approach because the estimation of demand and supply curves is less straightforward than in Chap. 1 because the VTR targets vehicles, thus affecting many individuals and industries (although logistics is plausibly the primary industry concerned) that use vehicles, and because the timing of compliance substantially differs across vehicle owners. Hence, rather than using a supply

---

[1] The official name of the law is the "Automobile Nitrogen Oxides–Particulate Matter Act."

curve, we estimate the cost by summing the compliance costs spent on all regulated vehicles.

The remainder of the chapter is organized as follows. Section 2.2 provides an overview and background of the VTR. Section 2.3 details the compliance methods that are adopted by vehicle owners. We discuss the estimation method and the results in Sects. 2.4 and 2.5. Section 2.4 focuses on the costs of the VTR, and Sect. 2.5 explores its benefits. Based on the results provided in these sections, we conduct a cost-benefit analysis in Sect. 2.6. Finally, Sect. 2.7 concludes the chapter.

2.2 A Brief History of Japanese Air Pollution Regulation: Background of the VTR

Before the VTR was imposed, vehicle emissions were controlled by a regulation known as the emission standard (ES). The ES controls emissions from new vehicles by establishing an upper limit for emissions intensity, which refers to the amount of chemicals released by a vehicle per kilometer. The first standard was established in 1966, targeting carbon monoxide (CO) emissions. Standards for NO_x and hydrocarbon (HC) emissions were added in 1973,[2] followed by those for PM emissions in 1993. The ES standards were increasingly tightened thereafter (Fig. 2.1).

Because the ES applies only to new vehicles to be sold in the market, it could not control emissions from old, more polluting vehicles. To address this problem, the government introduced the Automobile NO_x Regulation Act (also known as "the NO_x Act")[3] in 1992. The Act directly regulated emissions from old vehicles by prohibiting their inspection and registration. That is, the NO_x Act specified terminal years for old vehicles and thereby promoted their earlier replacement.

Because heavy traffic and air pollution are far more serious problems in metropolitan areas, the NO_x Act targeted 196 municipalities in Tokyo, Saitama, Kanagawa, Chiba, Osaka, and Hyogo that were not achieving the ambient air quality standard. Hereafter, we refer to these areas as non-attainment areas. However, the Act was not particularly effective in reducing NO_x emissions. In 1998, for example, only 43 (36) percent of the roadside air pollution monitoring stations met the national ambient air quality standard for NO_x (PM).[4]

[2] This standard is known as "the 1973 emissions standard."

[3] The official name of the Act is "the Act concerning special measures for total emission reduction of Nitrogen Oxides from automobiles in specified areas." This Act targets "specified areas," which refer to metropolitan areas in which air pollution is much more severe than in other regions in Japan and in which the ambient air quality standard was not met (non-attainment areas).

[4] According to the Environmental Information Center (EIC), roadside air pollution monitoring stations are defined as "monitoring stations established at places susceptible to automobile pollution, such as intersections, road and roadside, for the purpose of constantly observing the air quality (http://www.eic.or.jp/)."

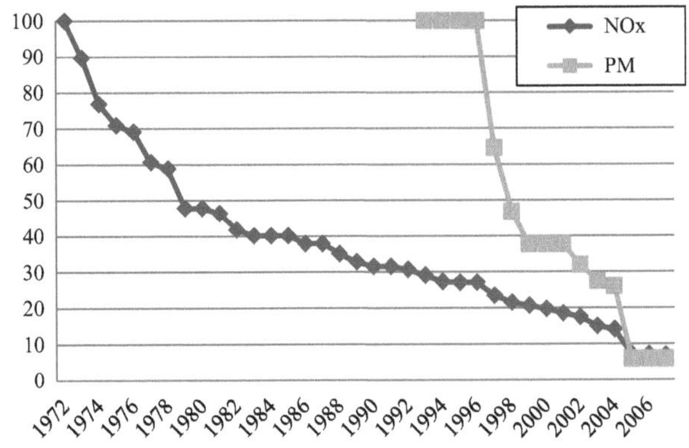

Fig. 2.1 Trend of tightening of NOx and PM emission intensities. Note: The values are averages because emission intensities are set by vehicle type. The standard values are set to 100. The standard values of NOx and PM are in 1972 and 1993, respectively

To reduce NO_x and PM[5] levels more effectively, the NO_x Act was replaced by the NO_x-PM Act[6] in 2001. The primary objective remains unchanged; the new Act also specifies terminal years for old vehicles to promote their earlier replacement. The NO_x-PM Act differs from the NO_x Act in that, as the name implies, the new act aims to decrease PM emissions as well as NO_x emissions, and its enforcement areas are extended to 276 municipalities, adding Nagoya to the original 196 municipalities. These areas, which are called "specially designated regions," are illustrated in Fig. 2.2.

Under the NO_x-PM Act, the VTR targets all old vehicles registered for parking in the enforcement areas in accordance with the Act on the Assurance of Car Parking Spaces and Other Matters. The VTR is the first attempt in the nation to reduce emissions from old vehicles. Regulations of this type are rare in other countries; typically, environmental regulations apply only to new vehicles, not to old vehicles already in use. The following vehicles are exempt from the VTR[7]: (1) gasoline passenger vehicles and special-use vehicles, (2) vehicles that satisfy the emissions standards for 2001 (that is, the year in which the Act was imposed), and (3) new vehicles that have been sold since 2002.[8] Terminal years for vehicle types will be discussed in Sect. 2.4.

The VTR appears to have contributed to reducing both PM and NO_x emissions, as PM and NO_x levels have gradually declined since 2001, although the PM level was highest in that year (Fig. 2.3).

[5] Before the ES included the standard for PM in 1993, there was no regulation for controlling PM emissions.

[6] The official name of the Act is "the Act concerning special measures for total emission reduction of Nitrogen Oxides and Particulate Matter from automobiles in specified areas (non-attainment areas)." The Act was revised in June 2001.

[7] See the Ministry of Environment (2005) for further details on the enforcement areas and regulated vehicles.

[8] In 2001, "the 2005 emissions standards" were issued, prohibiting the use and sale of vehicles that do not meet the standards imposed in 2002 and thereafter.

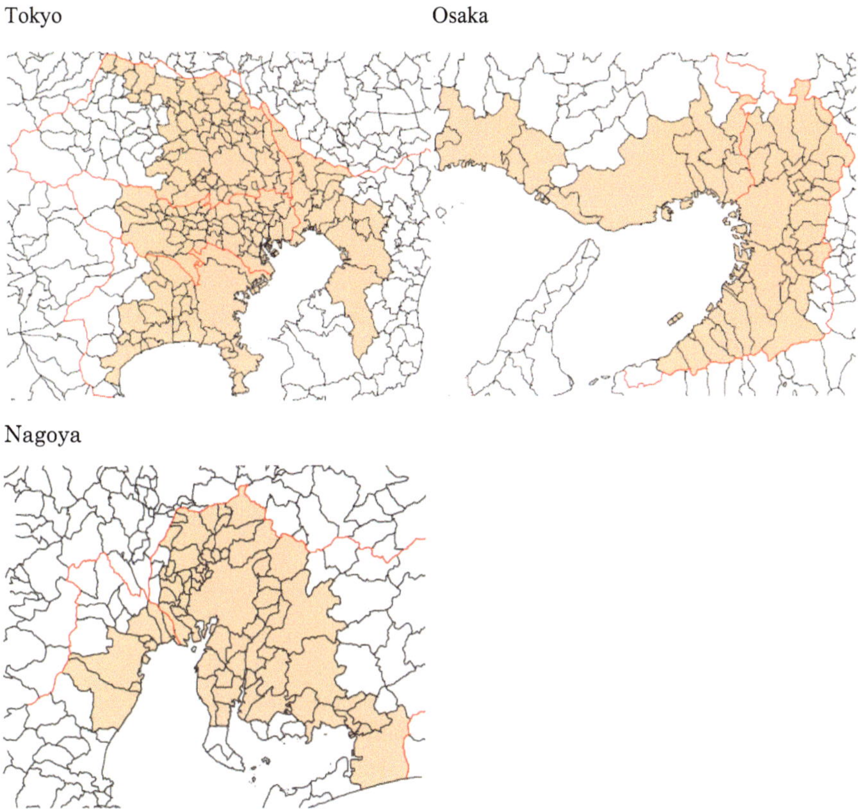

Fig. 2.2 The NO_x -PM Act: specially designated regions (From the Ministry of the Environment of Japan (2005))

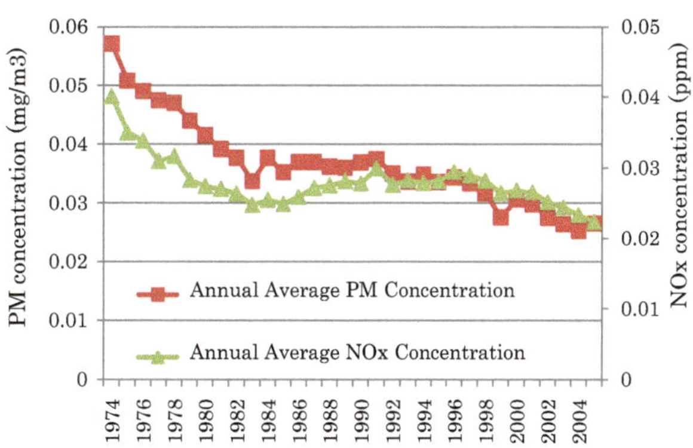

Fig. 2.3 Trend in NOx and PM concentration levels in specially designated regions

2.3 Compliance Methods

This section provides an overview of the methods for complying with the VTR. In particular, we focus on the methods used for diesel trucks because the VTR primarily targets diesel trucks, which had been regulated rather loosely under the ES. In this discussion, the user and owner of a truck will be assumed to be the same person. The effects and costs of the VTR may differ between small and standard diesel trucks. Thus, we will discuss the methods of compliance for each type of truck based on the Standard Truck Market Survey 2004 (2005a) and the Small Truck Market Survey 2004 (2005b) issued by the Japan Automobile Manufacturers Association (JAMA).

With regard to standard trucks, 71 % of the facilities that responded to the Standard Truck Market Survey 2004 responded that they were affected by the VTR. Replacement is the primary compliance method: 78 % of the facilities selected the response "replace with new vehicles that meet the VTR standards," and 14 % chose "replace with used vehicles that meet the VTR standards." In the enforcement areas, replacement occupies an even higher share: for example, 85 % of the facilities in Tokyo and 99 % in Osaka chose this method. Similar results were obtained for small trucks: more than 80 % of the facilities responded that they had replaced their vehicles with new trucks (JAMA 2005b).

In addition to diesel trucks, the VTR also regulates diesel passenger vehicles, buses, and special-use vehicles. Compliance methods for these vehicles are unclear, as they are not examined in the JAMA surveys or any other surveys. We thus assume that replacement is also the primary compliance method for these vehicles.

In the subsequent sections, we estimate the costs and benefits of the VTR by focusing on the case in which facilities replace older vehicles with new vehicles to comply with the VTR.[9]

2.4 Cost of the VTR

2.4.1 Compliance Cost

As mentioned previously, we have conducted a cost-benefit analysis by summing the compliance costs spent on all regulated vehicles. This method was chosen because the VTR is not a type of regulation for which the cost can be estimated using a straightforward supply curve or marginal cost.

If the cost of complying with the VTR increases the price of freight transportation, which in turn leads to a significant change in the demand for freight, then we would need to consider how consumer surplus changes. However, according to

[9] Replacement with used vehicles will be treated as replacement with new vehicles.

JAMA (2005a), the VTR and the NO_x-PM Act are generally unlikely to cause fare increases. Hence, in this study, the cost strictly refers to the cost of complying with the VTR. Under the VTR, vehicle owners are forced to take certain actions that they would not have otherwise taken. Therefore, the cost of complying with the VTR is the difference between the costs with and without the VTR.

The VTR was implemented in 2001 and actually began to be enforced in 2003. In the enforcement areas, vehicles not meeting the emissions standards are replaced during their terminal years. However, even in the absence of the VTR, old vehicles would be replaced with new vehicles at a certain point in time. The effect of the VTR is that the timing of replacement moves forward. That is, the VTR forces vehicle owners to sell their vehicles earlier than they would otherwise. If the opportunity cost of replacement is measured at the point of replacement, then it is equal to the difference between the following costs: (1) the cost of replacing vehicles at that point and (2) the cost of replacing vehicles without the VTR.

Vehicle owners sell their vehicles in the used car market earlier with the VTR than without the VTR.[10] Because younger vehicles have higher market values, vehicle owners gain additional income from selling their vehicles with the VTR than without the VTR. This difference in income can be adjusted by defining the cost of complying with the VTR as follows: the opportunity cost of accelerating the replacement timing as a result of the VTR subtracted by the profit from selling old vehicles.

2.4.2 Opportunity Cost of Replacement

In this section, we measure the opportunity cost of replacement per vehicle at the point of replacement. Let Y denote "reduced years," namely, the number of years that the VTR accelerates vehicle replacement. Without the VTR, the cost of purchasing a new vehicle is the present value of the vehicle that is discounted for period Y, as illustrated in Fig. 2.4. For example, on average, trucks that were initially registered in 1990 are replaced in 2010 in the absence of the VTR. By contrast, with the VTR, the same trucks must be replaced in 2005; thus, the number of reduced years is five. In the figure, "without" and "with" mean "without the VTR" and "with the VTR," respectively.

To measure the cost of replacement without the VTR in 2005, the price of a new vehicle must be discounted for the period of 5 years (2010–2005). Therefore, the replacement cost was computed as follows: (Price of a new vehicle) $\times$ exp ($-$discount rate $\times$ reduced years), where "exp $(\cdot)$" is an exponential function that discounts continuous variables. Let P denote the price of a new vehicle that is

[10] Because such vehicles are prohibited in Tokyo, Osaka, and Nagoya, they are sold and reused outside of these enforcement areas.

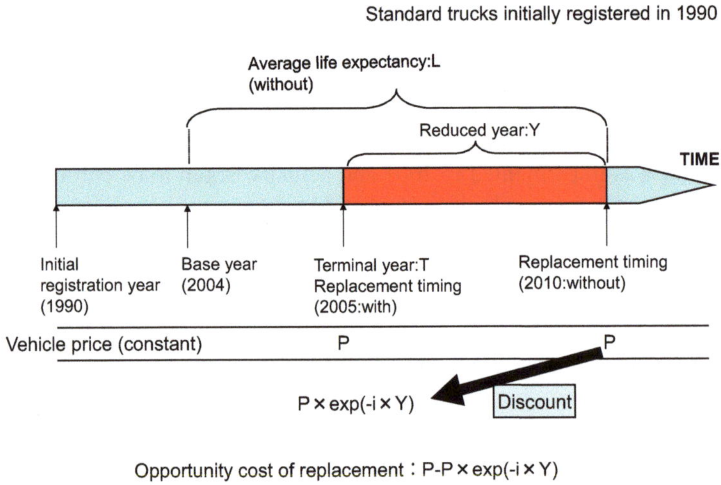

Fig. 2.4 Opportunity cost of replacement

constant over time, and i denotes the discount rate. Thus, the cost of replacement in 2005 is $P \times \exp(-\mathrm{i} \times 5)$.

With the VTR, the cost of replacement in 2005 is simply P. Because the compliance cost is the difference between the costs of replacement with and without the VTR, it is expressed as follows:

$$\begin{aligned}
&(\text{Opportunity cost of replacement})\\
&= (\text{Purchase cost with the VTR}) - (\text{Purchase cost without the VTR})\\
&= P - (\text{Present value of } P \text{ discounted for the period } Y)\\
&= P - \{P \times \exp(-\text{ discount rate} \times Y)\}
\end{aligned} \tag{2.1}$$

According to JAMA (2005a, b), vehicle owners do not change the types of vehicles upon replacement. Thus, P is assumed to be identical with or without the VTR.

To obtain Y, the following items must be computed for each vehicle type: (1) the point at which a vehicle is prohibited from use and (2) the average life expectancy of the vehicle at that point. With regard to the second item, the value prior to the implementation of the VTR must be used. Otherwise, the pure effect of the VTR cannot be measured because the VTR mandates that vehicles be retired at certain points, thereby shortening their life expectancies. Terminal years depend on vehicle types and initial registration years. Likewise, average life expectancies depend on vehicle types. To ensure the accuracy of Y, it is necessary to obtain as much detailed information on (2.1) and (2.2) as possible.

To compute (2.1), we first identified the terminal year for each vehicle type and initial registration year. Given that the registration is effective for 2 years, new standard trucks that were purchased and initially registered in 1990 are prohibited from use between October 1, 2004 and September 30, 2006 (Ministry of the Environment 2005). Given the assumption that initial registrations are uniformly distributed across the 12 months of 1990, these vehicles were retired in 2005 on average. We assume the same for vehicles that were initially registered in other years. That is, for each vehicle, we use the midpoint of the beginning and end of its terminal year. Table 2.1 presents the terminal year T for each vehicle type by initial registration year. For example, as shown in the table, standard trucks that were purchased in 1995 were terminated in 2006.

With regard to (2.2), we applied the method proposed by Oka et al. (2007). Using data in 2000, we computed average life expectancies by vehicle types (see Appendix 2.1 for our computational methodology). In principle, we could have used any other year, but this year was chosen because of data availability and because more recent data are preferable. This preference is especially notable given that the life expectancies of trucks have increased compared to those in 1990, for example, because of technological improvements and the deregulation of vehicle inspection (Oka et al. 2007). However, we must also choose from data in the years before the VTR was imposed; as mentioned above, life expectancies are likely to have been shortened under the VTR. As in the work of Oka et al. (2007), we assumed that vehicles that are used for 21 years are disposed of in the following year. That is, the average life expectancy was assumed to be 1 year for vehicles that are 21 years old. Based on these assumptions, we calculated the average life expectancy by vehicle type and age (see Appendix 2.1).

The average life expectancy at the point when a given vehicle is prohibited from use can be obtained by examining Table 2.1 and Appendix 2.1, in which the terminal years and average life expectancies are presented, respectively. For example, according to Table 2.1, standard trucks that were initially registered in 2002 were usable until 2012. Their reduced years Y are equal to the average life expectancy of a 10-year-old vehicle (2012–2002 = 10); thus, $Y = 7.13$. The values of Y[11] that were obtained in this manner are presented in Table 2.2 by vehicle type and initial registration year.[12]

Using Y, we formalize Cr^T below, the opportunity cost of replacing a vehicle that is prohibited from use in year T. Based on Eq. (2.1), Cr^T can be expressed as follows:

$$Cr^T = P[1 - \exp\{-i \times Y\}] \tag{2.2}$$

where P is the purchase cost.

[11] Computed based on terminal years and average life expectancy, Y is a function of the two.

[12] As shown in Table 2.2, Y for each vehicle type is fixed between 1997 and 2002 because during these years, the terminal year increased by one year if the initial registration year was one year later.

Table 2.1 Terminal years: *T*

IRY	STT	SMT	STB	SMB	STS	SMS	STP	SMP	STP	SM P
							Inspection every year		Inspection every two years	
2002	2012	2011	2015	2013	2013	2013	2012	2012	2012	2012
2001	2011	2010	2014	2012	2012	2012	2011	2011	2011	2011
2000	2010	2009	2013	2011	2011	2011	2010	2010	2010	2010
1999	2009	2008	2012	2010	2010	2010	2009	2009	2009	2009
1998	2008	2007	2011	2009	2009	2009	2008	2008	2008	2008
1997	2007	2006	2010	2008	2008	2008	2007	2007	2007	2007
1996	2006	2006	2009	2007	2007	2007	2006	2006	2006	2006
1995	2006	2006	2008	2006	2006	2006	2006	2006	2005	2005
1994	2006	2005	2007	2006	2006	2006	2006	2006	2005	2005
1993	2005	2005	2006	2006	2006	2006	2005	2005	2005	2005
1992	2005	2005	2006	2005	2005	2005	2005	2005	2005	2006
1991	2005	2005	2006	2005	2005	2005	2005	2005	2005	2005
1990	2005	2004	2005	2005	2005	2005	2005	2005	2005	2005
1989	2004	2004	2005	2005	2005	2005	2004	2004	2005	2005
Before 1988	2004	2004	2005	2004	2004	2004	2004	2004	2005	2005

Note: *IRY* initial registration year, *STT* standard trucks, *SMT* small trucks, *STB* standard buses, *SMB* small buses, *STS* standard special-use vehicles, *SMS* small special-use vehicles, *STP* standard passenger vehicles, and *SMP* small passenger vehicles

Table 2.2 Reduced years: *Y* (year)

IRY	STT	SMT	STB	SMB	STS	SMS	STP	SMP	STP	SMP
							Inspection every year		Inspection every two years	
1997–2002	7.13	5.93	5.69	5.25	6.61	5.25	5.94	4.10	5.94	4.10
1996	7.13	5.57	5.69	5.25	6.61	5.25	5.94	4.10	5.94	4.10
1995	6.61	5.25	5.69	5.25	6.61	5.25	5.25	3.55	5.94	4.10
1994	6.12	5.25	5.69	4.97	6.12	4.97	5.05	3.59	5.25	3.55
1993	6.12	4.97	5.69	4.72	5.69	4.72	5.05	3.59	5.05	3.59
1992	5.69	4.72	5.28	4.72	5.69	4.72	4.63	3.17	4.63	3.17
1991	5.28	4.47	4.84	4.47	5.28	4.47	4.66	3.40	4.66	3.40
1990	4.84	4.47	4.84	4.21	4.84	4.21	4.24	3.07	4.24	3.07
1989	4.84	4.21	4.35	3.90	4.35	3.90	4.24	3.07	4.16	3.27
1988	4.35	3.90	3.83	3.90	4.35	3.90	4.16	3.27	3.75	2.98
1987	3.83	3.54	3.25	3.54	3.83	3.54	3.75	2.98	3.38	2.85
1986	3.25	3.09	3.25	3.09	3.25	3.09	3.38	2.85	2.69	2.41
1985	2.61	2.54	2.61	2.54	2.61	2.54	2.69	2.41	1.89	1.79
1984	1.87	1.84	1.87	1.84	1.87	1.84	1.89	1.79	1.00	1.00

Note: *IRY* initial registration year, *STT* standard trucks, *SMT* small trucks, *STB* standard buses, *SMB* small buses, *STS* standard special-use vehicles, *SMS* small special-use vehicles, *STP* standard passenger vehicles, and *SMP* small passenger vehicles

2.4.3 Income from Selling Old Vehicles

We then formalized the income obtained from selling old vehicles. Because the price of a new vehicle is relatively higher, owners gain more income by selling them under the VTR than in its absence. To incorporate this relationship, we used the yearly average depreciation rate *sr* provided in the work of Kuroda et al. (1997); for example, the *sr* of a standard truck is reported as 25.72 %. Under the VTR, the income from selling a vehicle priced at P at its terminal year T is expressed as $P[\exp\{-sr \times (T - r)\}]$ where r is the initial registration year. Without the VTR, the income is $P[\exp\{-sr \times (T + Y - r)\} \times \exp\{-i \times Y\}]$. Therefore, the difference between the incomes with and without the VTR, Cs^T, measured at T is as follows:

$$\begin{aligned} Cs^T = P\big[&\exp\{-sr \times (T - r)\} \\ &- \exp\{-sr \times (T + Y - r)\} \times \exp\{-i \times Y\}\big] \end{aligned} \quad (2.3)$$

2.4.4 Compliance Cost

The cost of complying with the VTR for an additional vehicle, C^T, is the difference between Eqs. (2.2) and (2.3), i.e., between the opportunity cost of replacing it, Cr^T, and the income from selling it, Cs^T:

$$C^T = Cr^T - Cs^T \quad (2.4)$$

To calculate C^T, we must know the price of a vehicle, P, as Cr^T and Cs^T are functions of P. For the prices of trucks by load capacity, we obtained information from the Japan Trucking Association (2004). The average price of passenger vehicles was obtained per weight category based on data from the Japan Automobile Dealers Association (2000).

Based on the information above, we computed the compliance cost by initial registration year, vehicle type, and weight. We then averaged the costs across weights, thereby obtaining the cost for each combination of vehicle type and initial registration year. Because T depends on vehicle types and initial registration years, the compliance cost in the table was adjusted to the present discounted value in 2004 using a discount rate of 3 % as in Oka et al. (2007) (Table 2.3).

The table shows that even among vehicles of the same type, the average compliance cost differs greatly across initial registration years. For example, the average compliance cost for standard trucks is 438,000 yen if they were registered in 2000, but this cost decreases to 157,000 yen if the vehicle was registered in 2001. This dramatic decline occurred primarily because the types of vehicles that are regulated differ depending on the initial registration year. Among vehicles that were initially registered in 2000, the majority of those subject to the VTR are large and high-priced vehicles, whereas small vehicles had already met the VTR standards. Because new large vehicles had met the standards by 2001, the average compliance cost decreased that year.

Table 2.3 Average compliance cost per vehicle (10,000 yen/vehicle)

							STP	SMP	STP	SMP
IRY	STT	SMT	STB	SMB	STS	SMS	Inspection every year		Inspection every two years	
2002	15.8	11.3		10.4	14.1	11.4	13.7	3.4	13.7	3.4
2001	15.7	11.6	31.2	10.9	14.1	11.4	14.1	3.6	14.1	3.6
2000	43.8	11.8	35.4	11.6	25.3	11.6	14.5	3.7	14.5	3.7
1999	35.1	12.4	32.6	15.4	25.6	11.8	15.4	3.9	15.4	3.9
1998	33.7	12.7	32.9	16.1	25.9	12.1	15.7	4.0	15.7	4.0
1997	36.6	13.7	33.7	17.2	27.6	12.7	16.8	4.3	16.8	4.3
1996	38.9	14.8	36.0	18.0	26.7	13.4	17.2	4.3	17.2	4.3
1995	41.8	14.1	38.9	18.9	27.4	13.6	19.6	5.7	18.9	4.8
1994	38.6	15.1	44.0	19.4	27.2	13.4	21.5	6.9	20.2	5.5
1993	39.6	15.4	47.8	18.7	26.5	13.0	24.6	8.2	24.6	8.2
1992	38.2	15.3	43.9	21.4	32.2	14.4	22.9	6.9	22.9	6.9
1991	37.1	15.0	41.0	20.6	31.5	13.7	26.8	9.8	26.8	9.8
1990	34.4	17.3	46.2	20.0	28.4	13.6	23.5	7.6	23.5	7.6
1989	37.3	16.9	42.9	19.1	25.9	12.9	30.4	10.9	27.5	10.8
1988	34.5	16.3	37.3	21.8	30.7	15.3	28.9	10.2	23.4	8.4
1987	29.7	15.2	32.1	20.6	27.8	14.8	29.8	11.8	23.7	10.0
1986	24.4	13.7	38.2	18.6	23.7	13.6	27.1	10.8	18.8	7.9
1985	19.9	11.5	28.5	14.8	19.0	11.3	24.4	10.4	14.8	6.7
1984	16.3	7.9	23.0	9.7	13.8	7.6	17.9	7.9	8.3	3.8

Notes: Because none of the standard buses that were registered in 2002 were subject to the VTR, their cost is not reported in the table

IRY initial registration year, *STT* standard trucks, *SMT* small trucks, *STB* standard buses, *SMB* small buses, *STS* standard special-use vehicles, *SMS* small special-use vehicles, *STP* standard passenger vehicles, and *SMP* small passenger vehicles

2.4.5 *Total Cost*

The total cost can be obtained by summing C^T for all regulated vehicles. Let NR^t denote the number of vehicles replaced in year t as designated by the VTR. The total cost in year t is $C^t \times NR^t$. This cost, as evaluated in terms of the present discounted value in 2004, is $C^t \times NR^t \times \exp(-i \times (t - 2004))$. By summing this cost for all regulated vehicles, we can obtain TC, the total cost of complying with the VTR:

$$TC = \sum \sum_{t=2004} C^t \times NR^t \times \exp(-i \times (t - 2004)) \qquad (2.5)$$

where the first Σ represents summing over vehicle types and (t – 2004) reflects the fact that none of these vehicles have terminal years prior to 2004.[13]

To compute TC in Eq. (2.5), we needed to measure NR^t, the number of replaced vehicles, which in turn requires N^t, the number of old vehicles in year t (where t is some year after 2004). In this study, we first identified the number of regulated vehicles as of March 2003. Among the vehicles that were registered by 2003, some may have been replaced before their terminal years as a result of mechanical failures and traffic accidents. This "natural replacement" was removed from our estimates of N^t. See Appendix 2.2 for details on how we estimated N^t and NR^t.

The number of regulated vehicles in the enforcement areas was identified as follows. First, we obtained vehicle inspection data from the AIRIA (Automobile Inspection and Registration Information Association) database based on the following categories: region, types of vehicles surveyed in the database, weight, initial registration year, types of fuel surveyed in the database, and emissions code. A total of 3,917,553 vehicles were found to be registered in the enforcement areas, among which 2,594,949 vehicles are subject to the VTR (i.e., $N^{2003} = 2,594,949$). The second column in Table 2.4 presents N^{2003} by vehicle type. It should be noted here that one cannot determine whether a vehicle is subject to the VTR by simply observing its fuel type and initial registration year. For example, although trucks weighing more than 3.5 tons are permitted to be used if they meet the 1998 standards, those subject to the 1994 standards cannot be used. The VTR criteria differ depending on fuel type and weight. In addition, among the municipalities that belong to the same prefecture, some are included in the enforcement areas, whereas others are not. Hence, we carefully checked the emissions codes of the vehicles and the municipalities in which they are registered.[14]

Using a discount rate of 3 %, we computed TC and found a value of 521 billion yen in all enforcement areas. The third column in Table 2.4 presents the TC by vehicle type.

---

[13] See Table 2.1.

[14] Many vehicles in the data are missing emissions codes: 9 % (350,000 vehicles) of the vehicles in the enforcement areas do not have emissions codes. For these vehicles, we inferred the number of regulated vehicles based on the proportion of the regulated vehicles in the vehicles with emissions codes. We adopted this procedure for each combination of initial registration years, areas of registration, types of vehicles surveyed in the database, and fuel types. The VTR is more stringent for diesel passenger vehicles that are inspected every year (e.g., taxis and rental cars) than for those that are inspected every two years. However, we cannot distinguish between the two types of vehicles from the inspection data. Hence, we inferred the proportion of vehicles inspected every year on the basis of the proportion of diesel fuel used by commercial vehicles to that used by non-commercial ones. The amount of diesel fuel used in March 2003 is 4,391 kl by commercial vehicles and 489,514 kl by non-commercial vehicles (Monthly Statistical Report on Motor Vehicle Transportation 2003), and hence, 4,391/489,514.

Table 2.4 Number of regulated vehicles (N^t) and total cost (*TC*) (in hundreds of million yen)

Vehicle type		Number of vehicles	Cost
SMT		893,415	1,252.6
SMP	Inspection every two years	407,772	262.6
	Inspection every year	3,691	2.5
SMB		19,001	33.3
SMS		31,634	39.6
STT		581,192	2,112.0
STP	Inspection every two years	358,973	679.7
	Inspection every year	3,249	6.2
STB		27,638	107.8
STS		268,384	714.1
Total		2,594,949	5,210.4

Note: *SMT* small trucks, *SMP* small passenger vehicles, *SMB* small buses, *SMS* small special-use vehicles, *STT* standard trucks, *STP* standard passenger vehicles, and *STS* standard special-use vehicles

2.4.6 *Potential Biases in* TC

We should note that there are potential biases in our estimates of *TC*. Because fuel efficiency was expected to improve as a result of replacing older vehicles with newer vehicles, facilities that own many trucks may choose to use newer trucks with better fuel economy more frequently than old trucks (Nomura 2002). The replacement of old vehicles may also lead to reduced maintenance costs because maintenance costs and vehicle age are generally inversely related (Myojo et al. 2005). These considerations were not incorporated into our estimates because of data limitations. As such, we may have overestimated the true cost.

When replacing old vehicles, owners may purchase approved used vehicles rather than new vehicles. Because data on the prices of used trucks were not available when we conducted this study, we assumed that old trucks are exclusively replaced with new trucks. As the prices of used vehicles are lower than those of new vehicles, our model may again overestimate the true cost. However, because the share of used trucks is small (JAMA 2005a, b), we believe that the amount of bias from this source is limited.

Even after the use of certain vehicles is prohibited in the enforcement areas, such vehicles may be moved to another location and used elsewhere; registrations can be transferred to regions that are close to but outside of enforcement areas. In that case, the NO_x -PM Act along with the Act on the Assurance of Car Parking Spaces and Other Matters requires that owners update their addresses. This process can be particularly costly for owners of vehicles; hence, only a small number of firms choose to do so, as reported by Yokemoto (2007).

Finally, we estimated the share of commercial passenger vehicles based on their fuel consumption. Because commercial vehicles tend to drive longer distances than non-commercial vehicles, we may have overestimated the number of commercial vehicles, resulting in an overestimation of the true cost.

In summary, our estimates of the total cost are upwardly biased and are thus considered to be the upper bounds of the true costs. However, the errors in the estimates are likely to be small.

2.5 Benefit of the VTR

In this section, we explain how we computed the benefits of the VTR, which are measured in terms of reductions in negative externalities resulting from emissions reductions. Based on the procedure that we used to compute the cost of the VTR, we first estimated the volume of emissions reduction achieved by replacing each type of vehicle. We then summed the benefits from all vehicles to obtain the overall benefit of the VTR as shown below.

2.5.1 Effect of the VTR

The effect of the VTR is the difference between emissions in the absence of the VTR (i.e., the baseline) and emissions with the VTR, as depicted in Fig. 2.5. Let e_0^p (gram/litter) denote the latest emissions intensity of pollutant p for vehicles complying with the 2005 standards.[15] Likewise, let e_r^p denote the emissions intensity initially registered in year r—that is, the emissions intensity for older vehicles and, thus, $e_0^p < e_r^p$ for the same vehicle type, weight, and fuel type. Whereas the emissions standard for 3-ton standard trucks sold in 1990 is 0.154 g/l for PM ($e_{1990}^p = 0.154$), the same trucks sold in 2010 must meet the standard of 0.004 g/l for PM ($e_0^p = 0.004$), which is approximately 40 times more stringent than it was 20 years ago.

The shaded part of Fig. 2.5 is the volume of emissions reduction achieved by the VTR. Without the VTR, three-ton trucks sold in 1990 would have been used until 2010 (i.e., the end of their average life expectancy) and then replaced with new trucks complying with the latest stringent standard $\left(e_0^p = 0.004\right)$. Hence, even in the absence of the VTR, the amount of emissions decreases because the emissions standard for new trucks becomes more stringent. However, this reduction in emissions must be distinguished from the reductions achieved by the VTR. Under the VTR, the same trucks must be replaced in 2005. This earlier replacement leads to further emissions reduction, which is the effect of the VTR that we have attempted to measure.

[15] Although ES is expressed in gram/km, we use the unit gram/litter in Figure 2.5, assuming the average vehicle speed of 20 km/h.

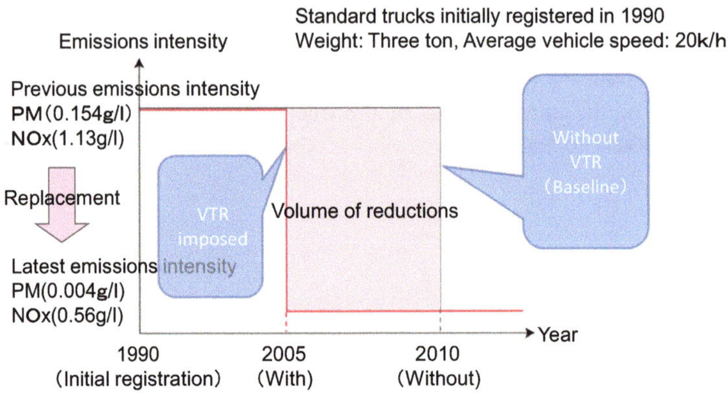

Fig. 2.5 Emissions reduction achieved by the VTR

2.5.2 Baseline: Emissions Without the VTR

Using the equation below, we computed emissions by vehicle type and regulated year:

$$\text{Emissions}(g) = (\text{Emissions Intensity})\,(g/km) \times (\text{Mileage})\,(km/vehicle) \times (\text{Number of Vehicles}) \tag{2.6}$$

The effect of the VTR on emissions reduction is, as noted above, the difference between emissions with and without the VTR.

We first computed the baseline, namely, emissions without the VTR. E_{wo} denotes emissions without the VTR. Based on Eq. (2.6), $E_{wo,r}^{t,p}$, the emissions of pollutant p from vehicles registered in year r in year t can be expressed as follows:

$$E_{wo,r}^{t,p} = e_r^p \times D \times N_r^t + e_0^p \times D \times \left(N_r^{2003} - N_r^t\right) \tag{2.7}$$

where N_r^t is the number of old vehicles still in use in year t. Thus, the first term represents emissions from existing older vehicles. In contrast, the term $N_r^{2003} - N_r^t$ represents the number of older vehicles being replaced with new vehicles as a result of natural replacement. Thus, the second term represents emissions from newer vehicles. Here, we assumed that mileage per vehicle (D) is constant over time after 2003.

2.5.3 Emissions with the VTR

We then estimated emissions with the VTR. To comply with the VTR, the owner of a vehicle either replaces the vehicle or simply disposes of it. Therefore, the number of vehicles in the enforcement areas is likely to decrease after 2003.[16] Thus, we assumed that regulated vehicles are abandoned at a constant rate.

Emissions under the VTR, $E_{w,r}^{t,p}$, are defined in the following set of equations.

$$E_{w,r}^{t,p} = \begin{cases} \left[e_r^p \times N_r^t + e_0^p \times \left(N_r^{2003} - N_r^t\right) \times D\right], & if\, T_r > t \\ \left[e_0^p \times NR_r^{T_r} + e_0^p \times \left(N_r^{2003} - N_r^{T_r}\right) \times D \times dr_r^t\right], & if\, T_r \le t \end{cases} \quad (2.8)$$

See Appendix 2.3 for additional details. dr_r^t is the mileage adjustment rate per vehicle in year t.

The first equation is the volume of emissions before vehicles reach their terminal years, T, and is thus equivalent to Eq. (2.7). Recall that NR^T stands for the number of vehicles replaced by VTR. Thus, the second is the volume of emissions after T, that is, after old vehicles are completely replaced with new vehicles. Thus, the emissions intensity that is used in the equation is exclusively represented by e_0.

2.5.4 *Emissions Reduction*

Based on Eqs. (2.7) and (2.8), we can estimate the amount of emissions reduction. $ER_r^{t,p}$, the reduction of emissions of pollutant p in year t, is the difference between emissions under the VTR (Eq. (2.8)) and emissions without the VTR (Eq. (2.7)), as expressed below:

$$ER_r^{t,p} = E_{wo,r}^{t,p} - E_{w,r}^{t,p} \quad (2.9)$$

The total emissions reduction in year t is computed by summing Eq. (2.9) over the registration year, vehicle type, weight, fuel type, and region, expressed as follows:

$$TER^{t,p} = \sum \sum_r ER_r^{t,p} \quad (2.10)$$

where the first Σ represents summing over all types of vehicles in the enforcement areas.

We used data on mileage per vehicle from the Monthly Statistical Report on Motor Vehicle Transportation (2003) provided by the Ministry of Land, Infrastructure, Transport, and Tourism (Japanese Department of Transportation). For the

[16] This assumption appears reasonable; the number of trucks that were registered in 2000 in Tokyo is 902,367, and this number declined to 791,391 in 2006 (http://www.airia.or.jp/). We also assumed that the VTR does not affect traffic demand. If we assume that mileage per vehicle is constant over time, then the total mileage decreases because the number of vehicles decreases. Hence, we imposed a restriction in which total mileage is constant in the presence of the VTR such that mileage per vehicle increases when the number of vehicles decreases.

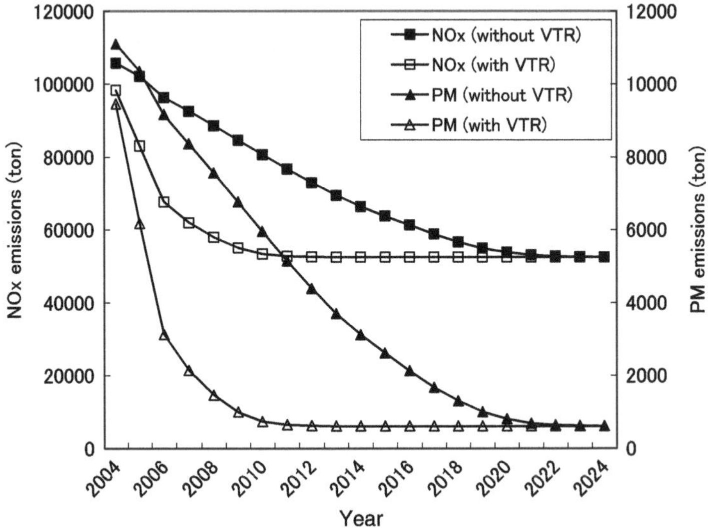

Fig. 2.6 Changes in the volume of NO_x and PM emissions: with and without the VTR

municipalities that are not included in the report, we used the average mileage per vehicle for each vehicle type. With regard to emissions factors, we used the results of a survey (Suri-Keikau Inc. 2005) and obtained NO_x and PM emissions intensities for the following categories: fuel type, vehicle type, years in which vehicles met the emissions standards, and speed per kilometer.

Using Eqs. (2.7), (2.8), and (2.9), we estimated the effect of emissions reduction. We used an average emissions intensity of 20 km per hour because it is approximately the average vehicle speed in Tokyo.

Figure 2.6 illustrates changes in the volume of NO_x and PM emissions under the VTR (Eq. (2.8)) and absent the VTR (Eq. (2.7)). By 2024, the total volume of NO_x (PM) emission will be 1,495,496 tons (87,624 tons) without the VTR and 1,212,319 tons (32,655 tons) with the VTR. That is, by 2024, the VTR reduces NO_x emissions approximately by 19 % and reduces PM emissions by 63 %.

As shown in Fig. 2.6, PM emissions decrease more rapidly than NO_x emissions. PM emissions intensity improved substantially because the emission standard became stringent more rapidly since 1993 compared with NO_x unit regulation (see Fig. 2.1).

When measuring the volume of emissions reduction, we did not account for compliance methods other than replacement, such as moving/registering regulated vehicles outside of the enforcement areas and reducing the number of vehicles owned. Hence, the effect of emissions reduction refers only to the effect of replacement.

Our estimate does not incorporate the possibility that the total number of vehicles changes after 2004; depending on economic conditions, firms may either increase or decrease the number of new vehicles that they purchase. We assumed that without the VTR, the total number of vehicles would be constant after 2003.

That is, the number of vehicles subject to the VTR as of 2003, which is approximately 2,590,000 vehicles, would remain unchanged after 2004. Note, however, that the number decreased under the VTR because some owners choose to discard their vehicles as a means of compliance.

2.5.5 Net Benefit

To compare the total costs and benefits of the VTR, we converted emissions reductions into a monetary value by multiplying NO_x and PM reductions by their respective marginal external cost (MEC). Let MEC^p represent the average MEC for pollutant p (NO_x or PM). The total benefit by 2004 is then expressed as follows:

$$TB = \sum \sum_t \sum_p \lfloor TER^{t,p} \times MEC^p \times \exp\{-i \times (t - 2004)\} \rfloor \qquad (2.11)$$

The first Σ indicates that emissions reductions per vehicle type are summed across all types of vehicles in the enforcement areas.

Table 2.5 presents the MEC estimates that we used in our analysis. We used estimates from the European Union (NETCEN) (2002) and those in the work of Koyama and Kishimoto (2001) for the benefits of NO_x reduction and PM reduction, respectively.

Estimates of MEC are known to be uncertain, as the causal relationship between pollutants and health has not been scientifically demonstrated in some cases. Thus, we conducted a sensitivity analysis to account for this uncertainty. The objective of this analysis was to examine how the estimates change given a range of important yet uncertain parameters. We considered two patterns: the case in which the damage from pollutants is small (i.e., the lower bound) and the case in which the damage is large (i.e., the upper bound). We also provide the average of the two estimates.

NETCEN (2002) reports the MEC of NO_x in 2000 in 15 European Union countries. We used the mean of the 15 countries (4,200 euro) as the lower bound and used the highest value (8,200 euro), which is the MEC in France, as the upper bound. We used the average of the upper and lower bounds as the median estimates

Table 2.5 Marginal external cost (10,000 yen/ton)

	NO_x	PM
References	NETCEN (2002)	Koyama and Kishimoto (2001)
Upper	MEC in France	Upper estimates
Median	Median value	Median estimates
Lower	Mean of 15 EU countries	Lower estimates
Upper	76.9	3,192.6
Median	58.1	2,276.3
Lower	39.4	1,360.0

for the MEC (i.e., 6,200 euro). These values were converted to Japanese yen in 2004 based on the exchange rate and GDP deflator in 2000.

We used the upper and lower bounds found in the work of Koyama and Kishimoto (2001), who estimated the MEC of air pollution, primarily that of PM, in Japan. Similar to the procedure used for NO_x, we used the average of the upper and lower bounds as the median estimates. We refer to the term "highest benefit" when using the upper bounds of both NO_x and PM, "medium benefit" when using the median estimates, and "lowest benefit" when using the lower bounds.

2.6 Cost-Benefit Analysis

To compare the total costs and benefits of the VTR, we converted them into monetary values. Table 2.6 presents the results. The total benefit is 1,202.2 billion yen using the median estimate. The net benefit, 681.2 billion yen, was obtained by subtracting the total cost from the total benefit.

Next, we present the results of the sensitivity analysis. The total benefit is 1,675.1 billion yen at the highest and 729.4 billion yen at the lowest. The finding that the highest benefit is more than double the lowest benefit results from the difference between the upper and lower bounds of the MEC. The increase in the net benefit is 1,154 billion yen at the highest and 208.3 billion yen at the lowest. Both values are positive, confirming the effectiveness of the VTR.

Our estimates of the benefit should be interpreted with caution, as they are likely to be biased. However, the direction of the bias can be inferred and is likely to be downward for the following reasons. First, both NETCEN (2002) and Koyama and Kishimoto (2001) estimated the MEC in terms of the medical expenditure, thus excluding disutility from discomfort. This method results in an underestimation of the MEC. Second, even if the MECs of NO_x and PM are precisely estimated, the MEC in the enforcement areas is unlikely to be near the lower bound. The probability is low because the lower bound of NO_x is the mean of the 15 EU countries, for which the average income and population values are smaller than those in the VTR enforcement areas in Japan; the VTR is imposed only in metropolitan areas in which the population is highly dense and the average income is the highest in the country. In addition, given that the total cost estimated in Sect. 2.4 is likely to be upwardly biased, the net benefit is presumed to be positive.

In sum, our estimates of the net benefit are downward biased; thus, the biases do not influence the discussion of the effectiveness of the VTR.

Table 2.6 Total cost and benefit: median estimate (in hundreds of million yen)

Total cost			5,210.4
Total benefit	Median value		12,022.2
		NO_x	1,388.7
		PM	10,633.5
Net benefit	Median value		6,811.7

2.7 Efficiency of the VTR: Simulation of Alternative Policies

2.7.1 *Marginal Abatement Cost*

Our cost-benefit analysis confirmed that the VTR provides benefits that substantially outweigh its costs. We then needed to consider whether the regulator set the terminal years from a cost minimization perspective (i.e., considering the differences in the MEC among the vehicle types). From the perspective of the equimarginal principle, if the MEC of NO_x and PM are not uniform across polluters, then the VTR has not been efficiently designed. In other words, if the VTR had been implemented more efficiently based on an ex-ante policy evaluation, then it could have yielded a greater amount of social surplus. To explore this possibility, we quantitatively examined the extent to which social surplus improves by changing the current terminal years for the vehicle types and initial registration years.

If an emissions reduction target is achieved at the minimum cost, then the marginal abatement cost (MAC) of the pollutants should be equalized among polluters (equimarginal principle).[17] Thus, we simulated the MAC under the current VTR. In our simulation, polluters are the users/owners of vehicles that are subject to the VTR. The cost of the VTR was computed in Sect. 2.4 and presented in Table 2.3. As is evident in Suri-Keikau Inc. (2005), emissions intensity decreases as a result of replacing old vehicles. Dividing the cost of the VTR by the volume of emissions reduction achieved as a result of replacing vehicles, we can obtain the cost required to reduce one gram of each pollutant. This cost can be considered the approximate value of the MAC. Figure 2.7 illustrates the MAC of vehicles that were initially registered in 1990.

One gram of NO_x reduction costs 1.80 yen for standard trucks and 8.92 yen for passenger vehicles (both initially registered in 1990). The finding that the abatement cost is lower for standard trucks and higher for passenger vehicles suggests that greater NO_x reduction from standard trucks and less NO_x reduction from passenger vehicles are desirable. That is, the efficiency of the VTR improves by (1) accelerating the terminal years for standard trucks, (2) delaying the terminal years for passenger vehicles, or (3) adopting both strategies.

The difference in costs is more substantial for PM reduction: one gram of PM reduction costs 7.17 yen for standard trucks and 58.53 yen for passenger vehicles (both initially registered in 1990). The reduction cost for passenger vehicles is approximately eight times greater than that for standard trucks. As in NO_x reduction, the result suggests that PM emissions can be reduced more efficiently by accelerating terminal years for standard trucks, by delaying terminal years for passenger vehicles, or by opting for both strategies.

[17] See Kolstad (2010).

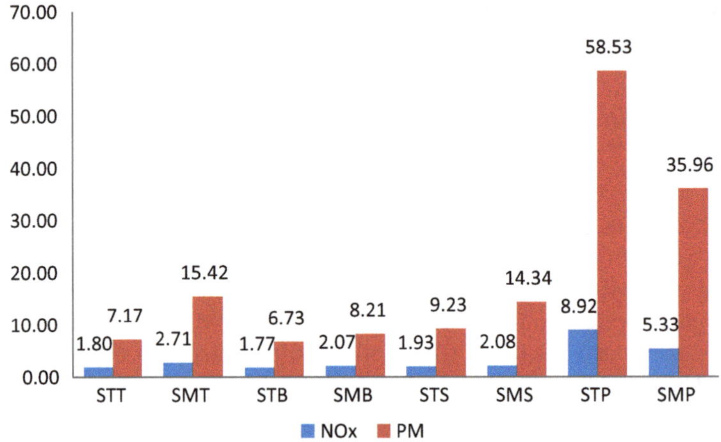

Fig. 2.7 Marginal abatement cost (compliance cost) of reducing PM and NO_x: initial registration 1990 (yen/g). Note: *SMT* small trucks, *SMP* small passenger vehicles, *SMB* small buses, *SMS* small special-use vehicles, *STT* standard trucks, *STP* standard passenger vehicles, and *STS* standard special-use vehicles

We then combined the environmental benefits of NO_x and PM reductions. In Table 2.7, we converted the reductions of these emissions into monetary values and computed the cost of reducing the externalities by one yen. The cost of reducing the negative externality by one yen is 0.32 yen for standard trucks and 2.44 yen for passenger vehicles (again, both are initially registered in 1990). The cost for passenger vehicles is nearly eight times larger than that for standard trucks, suggesting that for the VTR to operate more efficiently, terminal years should be advanced for standard trucks and shifted backward for passenger vehicles. The result also indicates the need to reconsider the efficiency of the VTR, particularly with regard to passenger vehicles that were initially registered in 1990; it is problematic that more than one yen is required to reduce the externalities generated by passenger vehicles by one yen.

2.7.2 *Simulation of Improved Efficiency*[18]

In the previous section, we showed that the MAC differs across polluters and that the efficiency of the VTR can be improved by changing the current terminal years. We then solved an optimization problem to maximize the net benefit by using the total cost denoted in Eq. (2.5) and the total benefit denoted in Eq. (2.11). We examined the extent to which the net benefit changes if we alter the policy variable,

[18] For additional details on the model in this section, see Iwata and Arimura (2009).

Table 2.7 Cost of reducing externalities by one yen

							STP	SMP	STP	SMP
IRY	STT	SMT	STB	SMB	STS	SMS	Inspection every year		Inspection every two years	
1984	0.26	0.69	0.25	0.30	0.31	0.57	3.47	2.13	3.45	2.01
1985	0.22	0.63	0.27	0.33	0.26	0.44	3.54	2.08	3.53	1.96
1986	0.21	0.60	0.26	0.32	0.26	0.45	3.46	1.99	3.39	1.84
1987	0.21	0.56	0.26	0.29	0.26	0.46	3.35	1.96	3.05	1.68
1988	0.21	0.54	0.25	0.28	0.27	0.46	2.98	1.93	2.78	1.77
1989	0.23	0.58	0.25	0.29	0.29	0.53	2.81	1.91	2.42	1.49
1990	0.32	0.64	0.30	0.36	0.40	0.59	2.44	1.49	2.44	1.49
1991	0.33	0.72	0.39	0.36	0.41	0.62	2.28	1.38	2.28	1.38
1992	0.32	0.69	0.38	0.34	0.40	0.57	1.85	1.02	1.85	1.02
1993	0.31	0.65	0.36	0.33	0.44	0.57	1.69	0.89	1.69	0.89
1994	0.30	0.59	0.36	0.30	0.44	0.49	1.34	0.55	1.71	0.82
1995	0.32	0.58	0.38	0.31	0.49	0.44	1.06	0.30	1.31	0.43
1996	0.35	0.47	0.49	0.36	0.51	0.43	1.04	0.27	1.04	0.27
1997	0.34	0.40	0.49	0.36	0.46	0.43	1.04	0.27	1.04	0.27
1998	0.35	0.42	0.49	0.35	0.46	0.46	1.09	0.27	1.09	0.27
1999	0.34	0.54	0.49	0.39	0.47	0.57	1.25	0.29	1.25	0.29
2000	0.32	0.67	0.49	0.60	0.51	0.73	1.39	0.30	1.39	0.30
2001	0.61	0.68	0.48	0.74	0.72	0.73	1.39	0.30	1.39	0.30
2002	0.63	0.67		0.73	0.99	0.70	1.39	0.30	1.39	0.30

Note: *IRY* initial registration year, *STT* standard trucks, *SMT* small trucks, *STB* standard buses, *SMB* small buses, *STS* standard special-use vehicles, *SMS* small special-use vehicles, *STP* standard passenger vehicles, and *SMP* small passenger vehicles

T (i.e., terminal years as presented in Table 2.1), for each vehicle type and each initial registration year.

Figure 2.8 illustrates the structure of this optimization problem. If *T* is altered, then the timing of replacement changes, which in turn changes *Y*. As a result, the cost of replacement and the income from selling old vehicles also change. This outcome further influences the number of vehicles to be replaced. In this manner, changing *T* results in a change in the total cost of the VTR.

Changing *T* also results in changes in the total benefit of the VTR in the following manner. First, a change in *T* influences the timing of the replacement of old vehicles with new vehicles, which naturally changes the timing of the adoption of the 2005 emissions standards at a given emissions intensity of e_0. This effect in turn changes E_w, the volume of emissions under the VTR, and then changes TER^t, the volume of emissions reduction achieved by the VTR in year t.

Table 2.8 presents the results when the net benefit is maximized,[19] specifically, the combination of *T* for each vehicle type and initial registration year under "the

[19] See Iwata and Arimura (2009) for further details on the maximization.

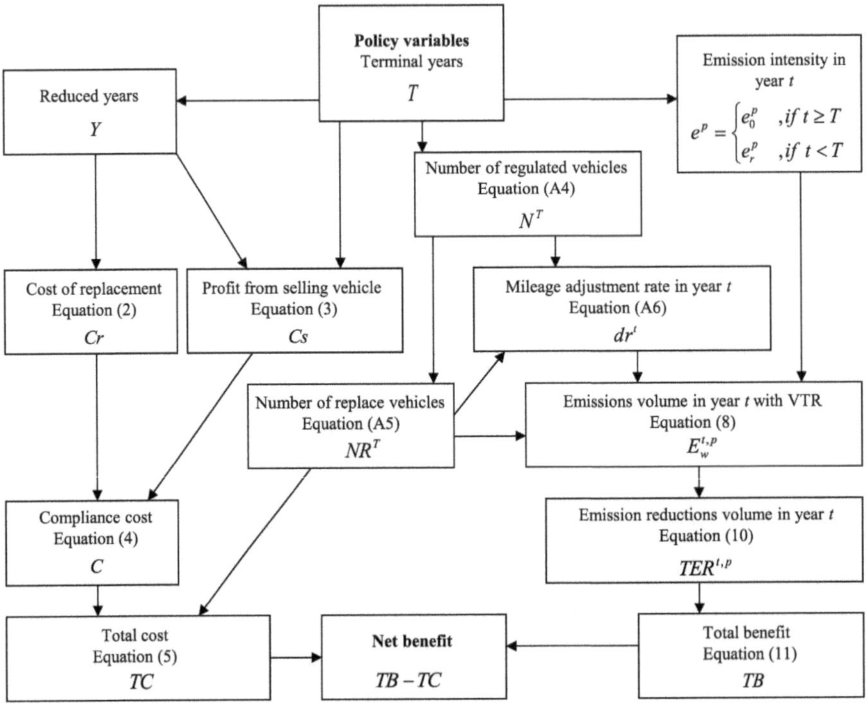

Fig. 2.8 Structure of optimizing simulation

optimal VTR" (median estimates). The optimal VTR is the solution to the optimization problem using the median estimates of MEC. As shown in the table, T is 2004 for all trucks. The emissions standard for trucks is not particularly stringent; hence, they produce a relatively high volume of pollution. The result suggests that the benefits would outweigh the costs if the VTR for trucks were strengthened.

In contrast, standard passenger vehicles that were initially registered in 1994 or before and small passenger vehicles that were initially registered in 1991 or before are no longer subject to the VTR. Passenger vehicles other than those were banned in 2004, as were similar trucks, because the benefit of discarding old vehicles is smaller than the cost, whereas new vehicles can be sold at high prices in used car markets.

Figures 2.9 and 2.10 presents the change in NO_X emissions volume (PM emissions volume) between the current and optimal versions of the VTR. The figures also illustrate how NO_X and PM emissions volumes change in the absence of the VTR. Even without the VTR, both NO_X and PM emissions volumes decrease over time as a result of natural replacement. Changes in emissions volume under the current VTR are indicated by the small black squares. As illustrated in the figures, both NO_X and PM emissions decrease more rapidly under the current VTR than in its absence. Compared with the current VTR, greater amounts of NO_X and

Table 2.8 Terminal years under the optimal VTR (median estimates)

							STP	SMP	STP	SMP
IRY	STT	SMT	STB	SMB	STS	SMS	Inspection every year		Inspection every two years	
2002	2004	2004	2004	2004	2004	2004	2004	2004	2004	2004
2001	2004	2004	2004	2004	2004	2004	2004	2004	2004	2004
2000	2004	2004	2004	2004	2004	2004	2004	2004	2004	2004
1999	2004	2004	2004	2004	2004	2004	2004	2004	2004	2004
1998	2004	2004	2004	2004	2004	2004	2004	2004	2004	2004
1997	2004	2004	2004	2004	2004	2004	2004	2004	2004	2004
1996	2004	2004	2004	2004	2004	2004	2004	2004	2004	2004
1995	2004	2004	2004	2004	2004	2004	2004	2004	2004	2004
1994	2004	2004	2004	2004	2004	2004	Excluded	2004	Excluded	2004
1993	2004	2004	2004	2004	2004	2004	Excluded	2004	Excluded	2004
1992	2004	2004	2004	2004	2004	2004	Excluded	2004	Excluded	2004
1991	2004	2004	2004	2004	2004	2004	Excluded	Excluded	Excluded	Excluded
1990	2004	2004	2004	2004	2004	2004	Excluded	Excluded	Excluded	Excluded
1989	2004	2004	2004	2004	2004	2004	Excluded	Excluded	Excluded	Excluded
1988	2004	2004	2004	2004	2004	2004	Excluded	Excluded	Excluded	Excluded
1987	2004	2004	2004	2004	2004	2004	Excluded	Excluded	Excluded	Excluded
1986	2004	2004	2004	2004	2004	2004	Excluded	Excluded	Excluded	Excluded
1985	2004	2004	2004	2004	2004	2004	Excluded	Excluded	Excluded	Excluded
1984	2004	2004	2004	2004	2004	2004	Excluded	Excluded	Excluded	Excluded

Note: *IRY* initial registration year, *STT* standard trucks, *SMT* small trucks, *STB* standard buses, *SMB* small buses, *STS* standard special-use vehicles, *SMS* small special-use vehicles, *STP* standard passenger vehicles, and *SMP* small passenger vehicles

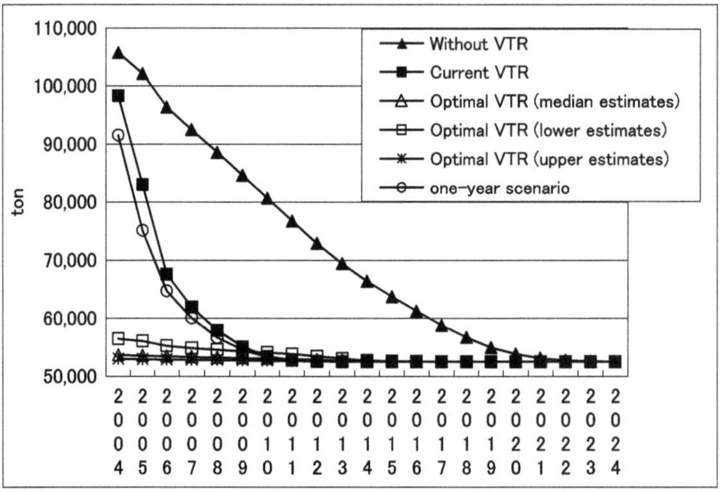

Fig. 2.9 Change in NO_x emissions volume under the optimal VTR

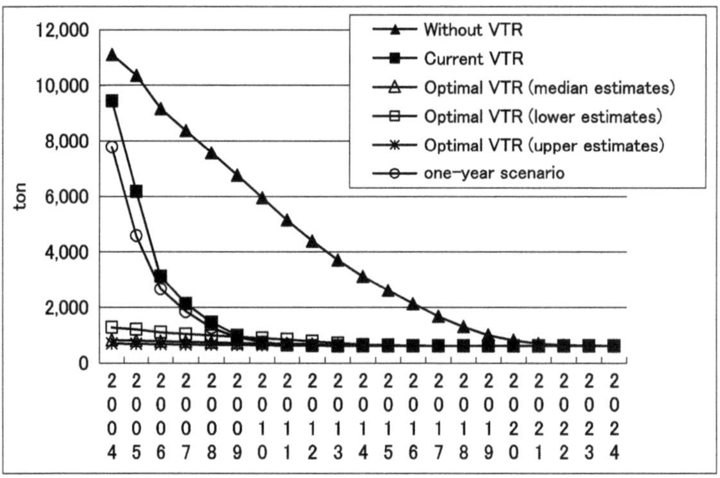

Fig. 2.10 Change in PM emissions volume under the optimal VTR

PM emissions are reduced under the optimal VTR. Because we assumed that vehicles 21 years old and above are discarded, the total emissions volumes become identical across all cases in 2024, when all vehicles are replaced with new vehicles.

Under the optimal VTR, the net benefit dramatically increases compared with the case with the current VTR. Table 2.9 presents the net benefit under the current VTR and that under the optimal VTR. This benefit is 681.2 billion yen under the current VTR and 1,388.4 billion yen under the optimal VTR. This result indicates that the net benefit could have increased by 104 % if the regulator had selected the most efficient terminal years.

Table 2.9 Net benefit under the current and optimal versions of the VTR (in hundreds of million yen)

	Estimates of the external cost		
	Median estimates	Lower estimates	Upper estimates
Current VTR	6,811	2,083	11,540
Optimal VTR	13,884	7,349	20,499
one-year scenario	7,722		

2.7.3 *Discussion on the Optimal Simulation*

In the previous section, we solved an optimization problem using median estimates. Because MEC estimates are known to be uncertain, we conducted a sensitivity analysis by solving the optimization problem with other MEC estimates. Specifically, we considered cases using the lower and upper bounds of the MECs (hereafter called "lower estimates" and "upper estimates," respectively) that are presented in Table 2.9.

Table 2.9 presents the increase in the net benefit using the lower and upper estimates along with the median estimates. As the table shows, the net benefit doubles irrespective of the MEC estimate that is employed. Hence, our results demonstrate the importance of an ex-ante policy evaluation to improve the efficiency of the VTR regardless of the size of the MEC. Figures 2.9 and 2.10 presents the changes in NO_x (PM) emissions volume according to the lower, median, and upper estimates.

Given all of the simulation results, we can robustly conclude that the current VTR is inefficient. To improve the efficiency of such regulations, an ex-ante quantitative analysis such as that performed in this chapter is necessary.

It is well known that the results of imposing environmental taxes would be identical to those of efficient regulation. Economic incentives are thus found to be superior to the current VTR.

Some may argue that the terminal years computed in our optimal simulation cannot be easily adopted, as they are drastically different from the current VTR. In response to this concern, we conducted a simple simulation in which the terminal years are changed only slightly from the current values. Specifically, the terminal years for passenger vehicles are delayed for 1 year, whereas those for standard trucks are accelerated by 1 year. No change was made for vehicles whose termination years are already 2004 under the current VTR. We refer to this situation as the "one-year scenario." The MEC that is used in this scenario is the median value. Even with these minor changes, the net benefit is found to increase by 13 % compared with that under the current VTR (see Table 2.9), and both NO_x and PM emission volumes were found to decrease slightly faster than under the current VTR (see Figs. 2.9 and 2.10). If the terminal years had been determined based on the results of an ex-ante policy evaluation, then minor and partial changes such as those in this scenario would have been possible, which could have resulted in a net benefit as large as 91 billion yen. Again, we would like to emphasize the importance of evaluating policies *before* they are actually implemented.

2.8 Conclusion

The NO_x-PM Act is a comprehensive regulation that specifies terminal years for 66 % of the vehicles in the enforcement areas. Using vehicle inspection data from AIRIA database, this study identified all vehicles subject to the VTR and computed the cost and benefit of the VTR in as much detail as possible. We estimated the compliance cost of the VTR (that is, the opportunity cost of earlier replacement) by vehicle type, initial registration year, and weight. After summing the results, we estimated the total cost to be 521 billion yen.

The benefit of the VTR was defined as the difference between the NO_x and PM emissions reduction volumes with and without the VTR. The VTR was found to drastically reduce both NO_x and PM emissions; the total NO_x and PM emissions reduction volumes were 283,000 and 55,000 tons, respectively. When converted into a monetary value, this reduction is equal to 1,202.2 billion yen.

One of the key elements of determining whether a policy should be supported is the amount of net benefit (i.e., the difference between the cost and benefit) that the policy delivers to society. In this study, the net social benefit of the VTR was estimated to be 681.2 billion yen. The benefit also remains positive in our sensitivity analysis, suggesting that the VTR is a reasonable policy from an economic perspective. However, this result does not necessarily indicate that the VTR has been implemented efficiently. In fact, we found that the MACs differ substantially across polluters. Using a simulation, we demonstrated that social surplus doubles if the current terminal years are changed to improve efficiency.

In this chapter, the VTR was found to be a legitimate policy by means of an ex-ante evaluation, although its legitimacy was unknown when it was implemented in June 2001. We suggest that regulators conduct cost-benefit analyses similar to that presented in this chapter to demonstrate the efficiency of a policy while it is being drafted. Finally, it should be noted that the results in this study showed that economic instruments such as emissions taxes are superior to prescriptive regulations. Economic theory predicts that the optimal regulation that is suggested in this study can be achieved by introducing an environmental tax. That is, by introducing an environmental tax, the government would have been able to implement efficient environmental policy without requiring much information.

Appendix 2.1 Average Life Expectancy: *L* (Year)

							STP	SMP	STP	SMP
Car age	STT	SMT	STB	SMB	STS	SMS	Inspection every year		Inspection every two years	
0	15.21	11.85	15.21	11.85	15.21	11.85	14.08	11.67	14.08	11.67
1	14.30	10.87	14.30	10.87	14.30	10.87	13.14	10.69	13.14	10.69
2	13.45	9.96	13.45	9.96	13.45	9.96	12.19	9.73	12.19	9.73
3	12.52	9.05	12.52	9.05	12.52	9.05	11.29	8.82	11.29	8.82
4	11.63	8.26	11.63	8.26	11.63	8.26	10.50	8.04	10.50	8.04
5	10.74	7.59	10.74	7.59	10.74	7.59	9.52	7.11	9.52	7.11
6	9.94	7.16	9.94	7.16	9.94	7.16	8.77	6.45	8.77	6.45
7	9.12	6.64	9.12	6.64	9.12	6.64	7.86	5.61	7.86	5.61
8	8.38	6.21	8.38	6.21	8.38	6.21	7.22	5.07	7.22	5.07
9	7.70	5.93	7.70	5.93	7.70	5.93	6.42	4.36	6.42	4.36
10	7.13	5.57	7.13	5.57	7.13	5.57	5.94	4.10	5.94	4.10
11	6.61	5.25	6.61	5.25	6.61	5.25	5.25	3.55	5.25	3.55
12	6.12	4.97	6.12	4.97	6.12	4.97	5.05	3.59	5.05	3.59
13	5.69	4.72	5.69	4.72	5.96	4.72	4.63	3.17	4.63	3.17
14	5.28	4.47	5.28	4.47	5.28	4.47	4.66	3.40	4.66	3.40
15	4.84	4.21	4.84	4.21	4.84	4.21	4.24	3.07	4.24	3.07
16	4.35	3.90	4.35	3.90	4.35	3.90	4.16	3.27	4.16	3.27
17	3.83	3.54	3.83	3.54	3.83	3.54	3.75	2.98	3.75	2.98
18	3.25	3.09	3.25	3.09	3.25	3.09	3.38	2.85	3.38	2.85
19	2.61	2.54	2.61	2.54	2.61	2.54	2.69	2.41	2.69	2.41
20	1.87	1.84	1.87	1.84	1.87	1.84	1.89	1.79	1.89	1.79
21	1.00	1.00	1.00	1.00	1.00	1.00	1.00	1.00	1.00	1.00

Note: *STT* standard trucks, *SMT* small trucks, *STB* standard buses, *SMB* small buses, *STS* standard special-use vehicles, *SMS* small special-use vehicles, *STP* standard passenger vehicles, and *SMP* small passenger vehicles

The average life expectancy L_m was calculated as follows. First, the number of registered vehicles of vehicle type m and vehicle age k, $N_m(k)$ was obtained from the "Survey on Vehicle Ownership in Japan" by AIRIA for standard trucks, small trucks, standard passenger cars, and small passenger cars. The disposal rate $d_m(k)$ was then computed as follows:

$$d_m(k) = (N_m(k) - N_m(k+1))/N_m(k) \tag{2.12}$$

From this disposal rate, the survival rate $s_m(k)$ was calculated:

$$\begin{aligned} s_m(k) &= s_m(k-1)\left[1-d_m(k-1)\right] && \text{if} \quad k \geq 1 \\ s_m(0) &= 1 && \text{if} \quad k = 0 \end{aligned} \tag{2.13}$$

Using the survival rate, the average life expectancy $L_m(A)$ for a vehicle of age A was then calculated:

$$L_m(A) = \frac{\sum_{k=A}^{21} s_m(k)}{s_m(k)} \tag{2.14}$$

Appendix 2.2 The Number of Vehicles Replaced, NR^t

Let N_r^{2003} represent the number of vehicles in 2003 that are initially registered in year r and $s(k)$ the survival rate calculated in (2.12). Using $s(k)$, we can compute N_{rm}^t, the number of vehicles of a given type in a given area in year t, as follows:

$$N_r^t = N_r^{2003} \times s(t-r)/s(2003-r) \tag{2.15}$$

where $t - r$ represents the vehicle age in year t and $2003 - r$ the vehicle age in 2003.

Using the above equation, we estimated the number of vehicles that are retired because of the VTR. For example, there were 581,192 regulated standard trucks in 2003 that were initially registered in 1990. Based on (2.15), by the time these trucks become subject to the ban in 2005, 70,949 of them would have already been replaced as a result of the natural replacement process. Thus, the number of vehicles actually subject to the ban in 2005 is 510,243.

Note that NR^t, the number of vehicles replaced, is not identical to the number of vehicles, N_r^t. According to JAMA (2005a, b), some vehicle owners choose to comply with the regulation not by replacing their vehicles but by disposing of them or using them outside of the enforcement areas. Thus, it is necessary to identify the extent to which regulated vehicles are replaced.

JAMA (2005a, b) conducted a firm-level survey in which firms are allowed to provide multiple answers. In cases in which a firm selects more than one practice for a given question, this study assumed that all of the chosen practices were employed with the same proportion. We also assumed that each firm has the same number of vehicles.

Based on these assumptions, we were able to compute the proportion of vehicles replaced as follows. First, we summed the percentage of each response and likewise summed the percentage of the "replacement" response. Dividing the latter by the former, we obtained the percentage of vehicles replaced. We found that the percentages of standard trucks replaced are 57 %, 71 %, and 42 % in Tokyo, Osaka, and Nagoya, respectively, and those for small trucks are 43 %, 68 %, and 61 % in Tokyo, Osaka, and Nagoya, respectively.

We use the percentage of small trucks for small buses and small special-use vehicles as well as that of standard trucks for standard special-use vehicles and standard buses for each region. Because information is not available on small and standard passenger vehicles and because these types of vehicles are unlikely to be moved outside of the enforcement areas for continuous use, we assumed that all of these vehicles are replaced (i.e., we assumed that their replacement percentage is 100 %).

Denoting rp as the replacement rate, we can express the number of vehicles replaced, NR_r^t, in year t for vehicle type m in region j without the regulation as follows:

$$NR_r^t = rp \times N_r^t \tag{2.16}$$

Appendix 2.3 Mileage Adjustment Rate

Because we assumed that $\overline{TD}_r$ is constant over time, dr_r^t varies subject to Eq. (2.17). The numbers of regulated vehicles after and before terminal year T_r are $N_r^{2003} - N_r^{T_r}$ $+NR_r^{T_r}$ and N_r^{2003}, respectively. Therefore, the first and second equations in (2.17) are the total mileages after T_r and before T_r, respectively.

$$\overline{TD}_r = \begin{cases} D \times dr_r^t \times \left(N_r^{2003} - N_r^{T_{rm}} + NR_r^{T_r}\right), & if\, T_r \leq t \\ D \times N_r^{2003}, & if\, T_r > t \end{cases} \tag{2.17}$$

References

Alberini A, Harrington W, McConnell V (1995) Determinants of participation in accelerated vehicle retirement programs. RAND J Econ 26(1):93–112

Alberini A, Harrington W, McConnell V (1996) Estimating an emissions supply function from accelerated vehicle retirement programs. Rev Econ Stat 78(2):251–265

Automobile Inspection and Registration Information Association (AIRIA) (1990–2006) Survey on vehicle ownership in Japan (in Japanese). AIRIA, Tokyo

Dill J (2004) Estimating emissions reductions from accelerated vehicle retirement programs. Transport Res Part D 9(2):87–106

Iwata K, Arimura TH (2009) Economic analysis of a Japanese air pollution regulation: an optimal retirement problem under vehicle type regulation in the NOx–particulate matter law. Transport Res Part D 14(3):157–167

Japan Automobile Manufacturers Association (JAMA) (2005a) Standard truck market survey 2004 (in Japanese). JAMA, Tokyo

JAMA (2005b) Small truck market survey 2004 (in Japanese). JAMA, Tokyo

Japan Automobile Dealers Association (2000) Annual report on new vehicle registration 23 (in Japanese). JADA, Tokyo

Japan Trucking Association (2004) Japanese trucking industry 2004. Japan Trucking Association, Tokyo, (in Japanese)

Kolstad CD (2010) Environmental economics, 2nd edn. Oxford University Press, New York

Koyama S, Kishimoto A (2001) External costs of road transport in Japan (in Japanese). Transport Policy Stud Rev 4(2):19–30

Kuroda M, Shinpo K, Nomura K, Kobayashi N (1997) KEIO database: measurement of output, labor and capital inputs. Keio economic observatory monograph series 8 (in Japanese). Tokyo

Lumbreras J, Valdés M, Borge R, Rodríguez ME (2008) Assessment of vehicle emissions projections in Madrid (Spain) from 2004 to 2012 considering several control strategies. Transport Res Part A 42(4):646–658

Ministry of Land, Infrastructure, Transport and Tourism (2003) Monthly statistical report on motor vehicle transportation (in Japanese). http://www.mlit.go.jp/k-toukei/search/pdf/06/06200303b00000.pdf. Accessed 15 May 2015

Ministry of the Environment (2005) Vehicle type regulation under the automobile NOx-PM Act (in Japanese). http://www.env.go.jp/air/car/pamph/. Accessed 15 May 2015

Myojo S, Kanazawa Y, Turnbull SJ (2005) Does TCO account for the U.S. consumer's preference for the automobile with Japanese nameplates? Proceedings of 2005 autumn conference of Japanese Economic Association. http://www2.chuo-u.ac.jp/econ/jeaf2005/program/docs/pm_18_7302a.pdf. Accessed 15 May 2015

National Environmental Technology Centre (2002) Estimates of the marginal external costs of air pollution in Europe. http://ec.europa.eu/environment/enveco/air/pdf/betaec02a.pdf. Accessed 15 May 2015

Nomura T (2002) A study on duration of standard trucks at facilities: effects on emission control. In: Kanemeto Y (ed) Study on control of CO_2 emission from transport sector (in Japanese). The Japan Research Center for Transport Policy, Tokyo, pp 26–51

Oka T, Fujii Y, Ishikawa M, Matsuo Y, Susami S (2007) Maximum abatement costs for calculating cost-effectiveness of green activities with multiple environmental effects. In: Huppes G, Ishikawa M (eds) Quantified eco-efficiency (eco-efficiency in industry and science). Springer, Dordrecht, pp 41–78

Suri-Keikau Inc. (2005) Vehicle emission intensities and total emission estimation survey 2004: commission research report to the ministry of the environment (in Japanese). Suri-Keikau Inc., Tokyo

Yokemoto M (2007) Environmental responsibility and cost sharing issues (in Japanese). Yuhikaku Publishing Inc., Tokyo

Chapter 3
Cost-Benefit Analysis of Enforcing Installation of Particulate Matter Elimination Devices on Diesel Trucks

Abstract In 2003, Japanese metropolitan municipalities in the Tokyo region introduced a unique regulation, called "operational regulation", to control air pollution from automobiles. The regulation requires old, dirty trucks and buses to install diesel particulate filters (DPFs) to eliminate particulate matter (PM) emissions. Regulated vehicles without DPFs are prohibited from running in the municipalities. However, there were no policy evaluations conducted before the legislation. This chapter examines a cost-benefit analysis of the regulation to determine whether it is a valid policy for obtaining cleaner air. The regulation is regarded as reducing health damage by mitigating emissions. The cost is the additional expenditure by vehicles' owners to comply with the regulation. As a result, the total cost and benefit are 36.1 and 264 billion yen, respectively. It is concluded that the operational regulation is a valid policy against air pollution because the net benefit is positive. Additionally, this result implies that the complementary policy concerning the replacement of old vehicles by new, cleaner ones is important and effective for further improving social welfare.

Keywords Air Pollution • Automobile • Regulation • Cost-benefit Analysis • Local Government • Particulate Matter

3.1 Introduction

To mitigate the problem of air pollution from automobiles, the government of Japan implemented an initial regulation, the emission standard (ES), in 1966. The standards for diesel trucks under the regulation were looser than those for gasoline vehicles, and the tax rate for diesel fuel was lower than that for gasoline (Hibiki and Arimura 2005). Furthermore, a lack of sufficient transportation infrastructure due to urbanization and motorization also made it hard to attain the recommended environmental air quality at road site air pollution stations in metropolitan areas. As a result, in the 1990s, a strong emphasis was placed on reducing concentrations of nitrogen oxide (NOx) and particulate matter (PM)[1] in these areas.

[1] This chapter deals with PM_{10} not $PM_{2.5}$ because there is no appropriate data on $PM_{2.5}$ in Japan.

T.H. Arimura, K. Iwata, *An Evaluation of Japanese Environmental Regulations*,
DOI 10.1007/978-94-017-9947-8_3

In response to the situation, the government of Japan and some municipality governments separately introduced different countermeasures against NOx and PM. In July 1992, the government of Japan introduced a new unique regulation, titled "Automobile NOx Act", which restricted the usage of old dirty diesel trucks in the Tokyo and Osaka metropolitan areas (this category of regulation is called "vehicle type regulation (VTR)"). The VTR requires use of old dirty vehicles to cease earlier than usual, setting terminal years.[2] However, it was pointed out that the effect of the regulation was limited, and in July 2001, the law was amended into the "Automobile NOx-PM Act" to make the regulation more effective (Sano 2008). The amended law regulated more areas and vehicles using the same regulatory method that the previous law had implemented.

After 2000, several municipality governments around Tokyo began to address the issue of air pollution from automobiles independently from the government of Japan. Initially, the "Tokyo Metropolitan Environmental Security Ordinance", which mitigated PM concentration, was approved at the council of Tokyo municipality in December 2000. The ordinance required old diesel trucks to use diesel particulate filters (DPFs), a new emission control technology (Lopez et al. 2009). Trucks without DPFs installed were prohibited from operating in the Tokyo municipality. This category of regulation is known as "operational regulation (OR)". Following the example of the Tokyo municipality, the same regulations were approved in the Tokyo region, that is, the neighboring municipalities (Saitama, Kanagawa and Chiba), between 2001 and 2002. Consequently, the OR started simultaneously in these municipalities in October 2003.

Regulators have several feasible policy options to reduce air pollution from automobiles, including prescriptive regulations or taxes on dirty fuel and vehicles (subsidies on clean ones). Policies, such as the famous corporate average fuel economy (CAFE)[3] in United States and the ES in Japan, are traditionally implemented to promote pollution or fuel efficiency control technologies for new vehicles. Therefore, many studies examine the effects of the CAFE. For example, Goldberg (1998) shows that the CAFE standard in 1989 generated a 201 million dollar loss on automakers and saved 400 million gallons because of giving them incentives to produce more fuel efficient vehicles. From a long-term point of view, Kleit (2004) and Austin and Dinan (2005) reveal that increases in the CAFE standards are less cost-effective than increases in the gasoline tax.

On the contrary, there are few policies to control emissions from old existing vehicles over the world. Because old vehicles are dirtier than the new vehicles because of technological innovation, it is important for regulators to pay attention to not only new vehicles but also old ones. The Japanese VTR and OR are examples of addressing the air pollution problem from old vehicles. Dill (2004) reveals that the accelerated vehicle retirement program introduced in California reduced air

[2] The terminal years refer to Chap. 2 of this book.

[3] The average fuel economy (mile per gallon) of vehicles produced by automakers is required to be less than the CAFE standard. If the average is over the standard, automakers must pay a fee.

pollutions. Lumbreras et al. (2008) estimates emissions including NOx and PM in Madrid from 2004 to 2010 under several scenarios, and they conclude that fleet renewal scenario is more efficient than the others. These studies may show the effectiveness of VTR-type regulation.

Although the methods of the VTR and OR are unique, the government of Japan and municipality governments did not conduct any ex ante policy evaluations on their regulations before passing the legislation. Arimura and Iwata (2008) examine cost-benefit analysis on the VTR and reveal that the VTR generates a positive net benefit to the order of 681.2 billion yen as shown in Chap. 2 of this book. Iwata and Arimura (2009) expand Arimura and Iwata's (2008) evaluation model and also conclude that the VTR is a valid policy for improving social welfare because the net benefit is positive. In addition, solving an optimal problem to maximize the social net benefit, they reveal that the net benefit can increase by 104 % if the regulator chooses optimal terminal years (Chap. 2).

As for the OR, there is no evidence from a cost-benefit analysis, similar to Arimura and Iwata (2008) and Iwata and Arimura (2009). Therefore, it has not been proved that the benefit of the OR would be more than the cost. Using cost-benefit analysis from an economics perspective, the tasks of this chapter are to quantitatively reveal both the costs and benefits of the regulation and to examine whether it is a valid clean air environmental policy. In addition, we present implications for future policy making against air pollution from automobiles to obtain further social welfare with comparing cost efficiencies between the VTR and OR.

The organization of this chapter is as follows: Section 3.2 explains the detailed differences between the OR and VTR. Sections 3.3 and 3.4 provide the models for calculating the cost and benefit from the OR, respectively. Section 3.5 presents the result of the cost-benefit analysis. A robustness check of the result is examined in Section 3.6. Section 3.7 is the conclusion.

3.2 Operational Regulation and Vehicle Type Regulation

OR and VTR have the same purpose: improving air quality in metropolitan areas. Their contents, however, are different in some respects, such as the regulated vehicle type[4] and the method of regulation.

Before viewing the differences, we must explain the automobile registration and inspection system in Japan.[5] After purchasing new vehicles, owners must perform two processes to use their vehicles. First, they must register their vehicles at the local District Land Transport Bureau under the Ministry of Land, Infrastructure, Transport and Tourism (Japanese Department of Transportation) in their living

[4] In this chapter, vehicles are divided into ten types. Only six types are regulated by the OR. The detail for the type is shown in Table 3.2.

[5] The system was established by the Road Transport Vehicle Act.

municipality. This initial registration is an important term in discussing the VTR and OR. Next, they must undergo the legal inspection of their automobile. If the installation of illegal car parts such as an altered muffler is found during the inspection, the vehicles are prohibited from operating on public roads. Vehicles must undergo this inspection every 2 years.[6] When a vehicle's owner changes or moves to other municipality, reregistration must be performed. The reregistration must take place in the new municipality. Unlike reregistration, reinspection is needed every 2 years even when owners change.

In 2002, the VTR became effective under the Automobile NOx-PM Act in Tokyo, Osaka and Nagoya, which are the biggest three metropolitan regions. Except for gasoline, hybrid and LPG passenger vehicles, all vehicles registered in the three regions with exhaust that does not meet the 2005 standard[7] are subject to the regulation. By setting banned years by vehicle type and initial registration year, the regulation promoted early replacement of dirty old vehicles by clean new ones. This regulation is constructed to prevent violations through its link to the described automobile registration and inspection system. That is to say, no owner can use his or her vehicle in violation of the regulation because the regulated vehicles are not eligible to undergo legal inspection by the VTR after the banned year. Although the dirty regulated vehicles registered in the regulated regions cannot operate after the year, if the owners sell them in the non-regulated region, new owners living there can buy and use them after undergoing reregistration.

The vehicles regulated under the OR are given a 7-year grace period before being subject to enforcement. After this period ends, unless a DPF that the municipalities have designated as an approved device is installed, a regulated vehicle cannot be used in the Tokyo region. Though a DPF's ability to eliminate PM from exhaust gas depends on the vehicle's type, weight and so on, the rate of elimination must be between 30 and 70 % (8Capital Pref City Aozora Network 2007). Unlike the VTR, only diesel trucks, buses and special-use vehicles with lower standards than the 1999 standard are regulated. Because the 1999 standard is looser than the 2005 standard, the OR regulates more dirty vehicles than the VTR does. The OR is not linked to the automobile registration and inspection system because it is based on municipal ordinances. Not only vehicles registered in the Tokyo region but also inflow vehicles from outside the Tokyo region are regulated. If a violation is found during spot checks on public roads by municipal officers, the owner is prohibited from using the vehicle on public roads or must pay a fine of less than 500,000 yen. Table 3.1 presents the differences between the two regulations.

Two remarkable points must be mentioned before showing the evaluation model. First, the OR is based on municipal ordinances, whereas the VTR is based on national law. Therefore, the two regulations are independent from each other,

[6] Commercial passenger vehicles, such as taxis and rental cars, must have this inspection once a year.

[7] The standards are set by vehicle type and weight under the ES. For example, the standard of diesel trucks with a weight of 2 tons is 0.63 g/km for NOx and 0.06 g/km for PM.

Table 3.1 Differences between the vehicle type regulation (VTR) and operational regulation (OR)

	Vehicle type regulation	Operational regulation
Administrator	Government of Japan	Municipalities
Launch	October 2002	October 2003
Purpose	Reduction of NOx and PM	Reduction of PM
Regulated area	Tokyo, Chiba, Saitama, Kanagawa, Aichi and Osaka municipalities (Tokyo, Osaka, Nagoya regions)	Tokyo, Chiba, Saitama and Kanagawa municipalities (Tokyo region)
Control subject 1	Vehicles not meeting 2005 emission intensity standard	Vehicles not meeting 1999 emission intensity standard
Control subject 2	All vehicles registered in the three regions except gasoline, hybrid and LPG passenger cars	Diesel trucks, buses and special-use vehicles registered in Tokyo region and inflow vehicles from outside Tokyo region
Content	Use prohibited after specific terminal years	Use prohibited in Tokyo region unless a diesel particulate filter is installed
Grace period	Designated years	7 years
Consequences of violation	Vehicle will not pass inspection	Use is forfeited or prohibited

and all vehicles are separated into the following three categories. There is no category in which vehicles must comply with the VTR but not with the OR, as the control vehicles of the OR are dirtier than that of the VTR.

Category (1) must comply with both regulations
Category (2) must comply with the VTR but not with the OR
Category (3) not required to comply with either regulation

The following are examples of these categories. Here, we assume that vehicles that are initially registered in year t satisfy the year t exhaust gas standard. Also, we assume that the year t standard is stricter than the year $t-1$ standard. As for diesel standard trucks registered in 1998, the VTR requires that their use be terminated in 2008 (Table 2.1 in Chap. 2). In addition, the OR requires that their owners purchase and install DPFs in them from 2005 to 2008 because the grace period of the OR is 7 years (Category 1). Diesel standard trucks registered in 2000 will be prohibited from running in 2010 by the VTR (Table 2.1 in Chap. 2), whereas they are not subject to the OR because they satisfy the 1999 standard (Category 2). Because diesel standard trucks registered in 2005 comply with both regulations, they have no obligations to follow them (Category 3). Category 3 also includes gasoline passenger vehicles.

Second, the effect of the VTR cannot be ignored when evaluating the OR because the VTR was legislated prior to the OR. Therefore, the benefits and costs of the OR are described as additional improvements in the air quality and social cost from the baseline established by the VTR.

Table 3.2 Number of regulated vehicles by regulation, municipality and vehicle type in the Tokyo region

Vehicle type regulation		Chiba	Kanagawa	Saitama	Tokyo	Total
Standard trucks		44,223	77,136	77,462	114,753	313,574
Small trucks		62,234	115,971	103,208	176,991	458,404
Standard buses		2,369	5,100	2,431	6,770	16,670
Small buses		1,681	2,859	3,006	3,585	11,131
Standard special-use vehicles		23,696	43,706	34,409	58,436	160,247
Small special-use vehicles		2,445	4,380	3,593	7,345	17,763
Standard passenger cars	Inspection: every 2 years	30,205	50,539	43,985	53,818	178,546
Small passenger cars		33,853	51,376	48,237	58,525	191,991
Standard passenger cars	Inspection: every 1 year	273	457	398	487	1,616
Small passenger cars		306	465	437	530	1,738
Total		**201,285**	**351,989**	**317,166**	**481,240**	**1,351,680**
Operational regulation						
Standard trucks		42,739	73,842	76,584	112,214	305,379
Small trucks		38,843	72,276	81,787	137,662	330,568
Standard buses		2,369	5,090	2,429	6,735	16,623
Small buses		1,566	2,470	2,951	3,371	10,358
Standard special-use vehicles		16,637	26,792	23,489	37,921	104,839
Small special-use vehicles		1,169	2,019	2,155	5,433	10,776
Total		**103,323**	**182,489**	**189,395**	**303,336**	**778,543**

Like Chap. 2, this chapter uses Survey of Automobile Possession (SAP) data from a legal automobile registration and inspection system provided by the Automobile Inspection and Registration Information Association (AIRIA). This dataset contains information about the characteristics of all vehicles in Japan, such as the vehicle type, the initial registration year, the fuel type, the weight, the registered location and the emission code.[8] Therefore, all regulated vehicles registered in the Tokyo region are identified, and this information is utilized in this chapter to perform a cost-benefit analysis of the OR.

In March 2003, there were 1.35 million vehicles regulated by the VTR in the Tokyo region. Of these, there were 0.78 and 0.57 million vehicles in Categories 1 and 2, respectively. The details of the regulated vehicles by vehicle type are presented in Table 3.2.

[8] The emission code is equivalent to emission intensity.

3.3 Cost of Operational Regulation

The total cost of the regulation is calculated by multiplying the compliance cost per vehicle by the number of regulated vehicles. We follow the basic structure of the model in Chap. 2. For instance, this chapter assumes that the owners of regulated vehicles are also their users. The cost is the difference between social costs with and without the regulation. Because owners in Categories 2 and 3 are unaffected by the regulation, only those in Category 1 incur some compliance cost.

How do owners in Category 1 comply with the regulation? It is safe to assume that they choose to install DPFs into their regulated vehicles (Method 1). In this case, the expenditure for purchasing the DPF is regarded as the cost. Note that they cannot avoid meeting the obligation of the VTR even when purchasing DPFs. Therefore, they must practice some additional compliance method for the VTR.

There is another option available, in which owners replace their old vehicles with new ones instead of installing DPFs (Method 2). Those owners who choose Method 2 do not need to use an additional compliance method for the VTR because new vehicles are not regulated by the VTR.

Japan Automobile Manufacturers Association reports (JAMA 2005a, b), which are surveys for the VTR, revealed that owners repurchased new vehicles of the same type as the old ones they used. On the contrary, no such survey exists in regard to the OR. Therefore, as in the VTR, owners are assumed to repurchase new vehicles of the same type they have used in the past when they choose Method 2 for complying with the OR. Table 3.3 indicates the compliance method combinations for both of the regulations.

3.3.1 *Cost of Method 1*

In Method 1, owners comply with the OR by purchasing DPFs. Therefore, the cost of the regulation is regarded as the expenditure for DPFs. Let *Pd* and *iny* be the expenditure (DPF price) and installation year, respectively. With a discount rate i, the cost per vehicle evaluated in 2004 can be calculated by Eq. (3.1).

Table 3.3 Compliance methods under each regulation

	Operational regulation	Vehicle type regulation
Method 1	Install DPF	Replace with new vehicles
		Partly not replace: rate (1-*rp*)
Method 2	Replace with new vehicles	Nothing
	Partly not replace: rate (1-*rp*)	

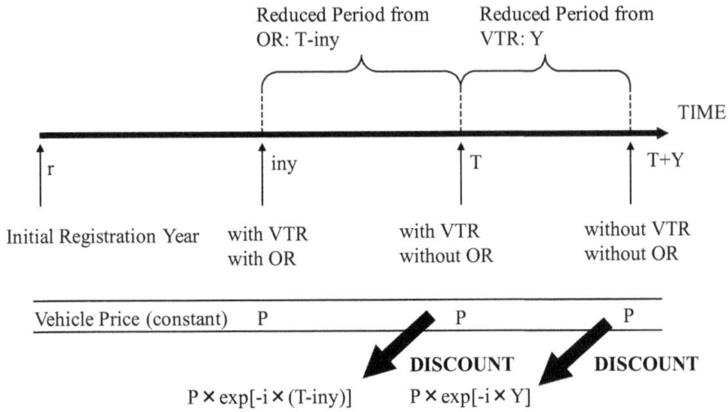

Fig. 3.1 Concept for cost per vehicle

$$CT1_r = Pd \times \exp[-i \times (iny_r - 2004)] \tag{3.1}$$

As the regulation gives owners seven grace years after the initial registration year r and takes effect after 2003, the installation year can be described by Eq. (3.2).

$$iny_r = \begin{cases} 2003 & ,if \;\; r + 7 \leq 2003 \\ r + 7 & ,if \;\; r + 7 > 2003 \end{cases} \tag{3.2}$$

3.3.2 *Cost of Method 2*

$CrT2$ is the cost per vehicle that is generated when owners choose Method 2 for the OR. Figure 3.1 illustrates the procedure for calculating the cost per vehicle. The calculation of the cost per vehicle for repurchasing new vehicles follows that of Chap. 2. Without the OR (and with the VTR), owners can use their old vehicles registered in year r until the banned year T_r, when they are prohibited from using them by the VTR. However, under the OR (and under the VTR), owners must replace old vehicles with new ones in the installation year iny_r. Therefore, vehicles' price P, the cost per vehicle evaluated in 2004 is an opportunity cost $CrT2_r$ due to early replacement, as shown in Eq. (3.3).

$$CrT2_r = [P - P \times \exp\{-i \times (T_r - iny_r)\}] \times \exp[-i \times (iny_r - 2004)] \tag{3.3}$$

In Fig. 3.1, $P - P \times \exp(-i \times Y_r)$ is the cost per vehicle for the VTR in banned year T_r.

Let sr represent a yearly average depreciation rate of vehicles. On one hand, without the OR, the residual value of the regulated vehicles in banned year T_r is $P - P \times \exp[-sr \times (T_r - r)]$. The term $T_r - r$ represents the vehicle's age in year T_r. On the other hand, the residual value in year iny_r is $P - P \times \exp[-sr \times (iny_r - r)]$ in the case of the OR. Therefore, the profit from selling old vehicles earlier, $CsT2$, as evaluated for 2004, is described by the following equation:

$$CsT2_r = P\left[\exp\{-sr \times (iny_r - r)\} - \exp\{-sr \times (T_r - r)\} \times \exp\{-i \times (T_r - iny_r)\}\right] \times \exp[-i \times (iny_r - 2004)]. \quad (3.4)$$

As a result, the cost of the OR, $CT2$, for Method 2 is defined as the difference between $CrT2$ and $CsT2$.

$$CT2_r = CrT2_r - CsT2_r \quad (3.5)$$

3.3.3 Total Cost

Which do the owners choose as a compliance method, Method 1 or 2? Based on cost-minimizing behavior, they must adopt the method with the lower cost. Therefore, the cost per vehicle CT is described by the following equation.

$$CT_r = \min[CT1_r, CT2_r] \quad (3.6)$$

The total cost is calculated by multiplying the cost per vehicle by the number of regulated vehicles. Before showing the total cost, we must explain how the number of regulated vehicles was obtained. N^{iny_r} denotes the number of regulated vehicles affected in year iny_r. Regarding the survival rate as s, $s(k)$ indicates the survival rate of k-year-old vehicle. Therefore, the number in iny_r can be calculated as in Eq. (3.7).

$$N_r^{iny_r} = N_r^{2003} \times s(iny_r - r)/s(2003 - r) \quad (3.7)$$

Equation (3.7) involves the effect of the natural reduction of vehicles without the regulation. Natural reduction means the replacement of old vehicles with new ones over time, even without regulation, due to accidents and so on. As this chapter uses a 2003 dataset, N_r^{2003} was obtained from the SAP data, similar to Chap. 2.

When owners choose Method 2, they repurchase new vehicles in year iny_r and then do not practice any compliance method in year T_r. JAMA (2005a, b), however, pointed out that some owners chose other compliance methods, such as reduction or disposal of their vehicles.

Therefore, this chapter takes these behaviors into account. Let $1 - rp$ be the rate of choosing to reduce or dispose of old vehicles (rp represents the rate of repurchasing new vehicles). Because there is no survey about this rate on the OR, we use the similar rate reported by Chap. 2 for the VTR as a substitute for it. The rate is also calculated by municipality and by vehicle type.

With repurchasing rate rp, the number of regulated vehicles registered in year r that will be replaced in year iny_r is calculated as follows:

$$NR_r^{iny_r} = rp \times N_r^{iny_r}. \tag{3.8}$$

All owners who choose Method 1 are assumed to purchase DPFs; that is, rp is zero because there is no evidence about the reducing/disposal rate for Method 1.

With the cost per vehicle CT, the numbers of regulated vehicles $N_r^{iny_r}$ and repurchased vehicles $NR_r^{iny_r}$, the total cost of the OR evaluated in 2004 is calculated in Eqs. (3.9) and (3.10):

$$TC = \sum \sum_r \left[CT_r \times \left\{N_r^{iny_r} \times I_r + NR_r^{iny_r} \times (1 - I_r)\right\}\right], \tag{3.9}$$

$$CT_r = \begin{cases} CT1_r & ,if\ I_r = 1 \\ CT2_r & ,if\ I_r = 0 \end{cases}, \tag{3.10}$$

where the first Σ represents summing over vehicle types and I_r is an indicator variable that equals one if owners whose regulated vehicles are registered in year r choose Method 1 and equals zero otherwise.[9] Therefore, in choosing Method 1, the number of vehicles denoted in Eq. (3.9) is $N_r^{iny_r}$ because the repurchasing rate is zero, whereas in choosing Method 2, the quantity is $NR_r^{iny_r}$ because the rate is not zero.

3.4 Benefit of Operational Regulation

The effect of the OR is defined as the reduction in air pollution emissions with versus without the regulation. Therefore, the benefit is a monetary value calculated from the reductions and subsequent improvements in public health.

The purpose of the OR is to decrease PM emissions through the enforcement of installing DPFs, whereas that of the VTR is to decrease both PM and NOx emission. In general, DPFs have the effect of eliminating only PM emissions but not NOx emissions. However, according to cost-minimization behavior, owners choosing Method 2 replace their old vehicles with new ones. New vehicles have good emission intensity not only for PM emissions (emission amount per kilometer) but also for NOx emissions. Therefore, to maintain a consistency of defined benefits between both regulations and to remain comparable with the result of Chap. 2, this chapter also handles PM and NOx emissions as the benefit.

[9] The number of classification for initial registration year (r) is 18, from 1985 (or before) to 2002.

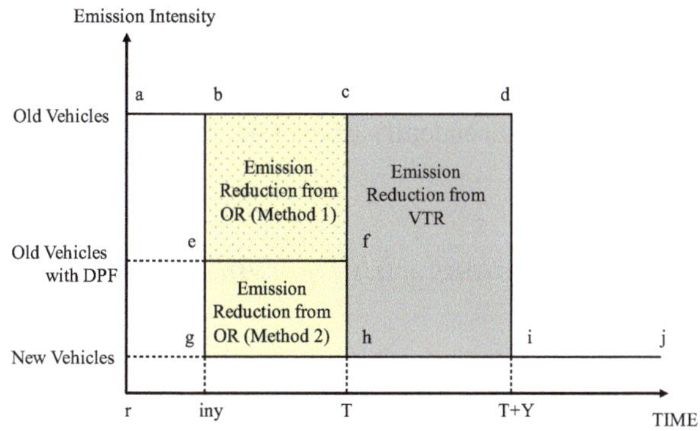

Fig. 3.2 Concept for emission reduction

3.4.1 Emissions Without Operational Regulation

Emissions are calculated by multiplying the emission intensity (ton/km) by mileage per vehicle (km) and by the number of vehicles. Figure 3.2 indicates a concept for calculating emission reduction similar to Fig. 3.1. The vehicles regulated by the OR will be banned in year T_r by the VTR, even without the OR. Therefore, because old vehicles are replaced by new ones in year T_r, the transition of emission intensity in Fig. 3.2 is a-b-c-f-h-i-j. Without the regulations, this order is described as a-b-c-d-i-j.

Let e_r^p be an old vehicle's emission intensity of pollutant p, and let e_0^p be the 2003 value. D represents the mileage per vehicle, which is assumed to be constant over time. Without the OR, yearly emission $E_{wo,r}^{t,p}$ in year t is expressed as Eq. (3.11)

$$E_{wo,r}^{t,p} = \begin{cases} \left[e_r^p \times N_r^t + e_0^p \times \left(N_r^{2003} - N_r^t\right)\right] \times D & ,if\ t < T_r \\ \left[e_0^p \times NR_r^{T_r} + e_0^p \times \left(N_r^{2003} - N_r^{T_r}\right)\right] \times dr_r^t \times D & ,if\ t \geq T_r \end{cases}. \quad (3.11)$$

The first and second equations in Eq. (3.11) are for yearly emissions before and after the banned year, T_r, respectively. In the first equation, the first term in parentheses represents emissions from old vehicles with an emission intensity e_r^p. Emissions from new replacement vehicles with a cleaner emission intensity e_0^p due to the natural reduction are described in the second term in parentheses.

The total number of vehicles is N_r^{2003} in the first equation but is not the same in the second. This difference results from the fact that the number of new replacement vehicles decreases to $NR_r^{iny_r}$ with precluding effects from the minor compliance methods, reducing/disposing vehicles, after T_r. The total number reduction after T_r decreases the total mileage because of the restriction of constant mileage per vehicle over time. It makes no sense that the total amount of mileage or cargo volume is significantly lessened only by the VTR. To tackle this problem, this chapter also incorporates an adjustment parameter dr_r^t under the assumption that the total mileage is constant over time, similar to Chap. 2.[10]

Equation (3.11) is equivalent to the emission equation of Chap. 2 because the VTR exists independently of the OR.

3.4.2 Emissions with Operational Regulation

With the regulation, the change of emission intensity depends on which compliance method owners choose. If owners choose Method 1, installing DPF, emission intensity decreases due to the installation in year iny_r. They then purchase new vehicles for the VTR while selling their old ones in year T_r. Therefore, in Fig. 3.2, the transition of emission intensity is represented as a-b-e-f-h-i-j.

If they choose Method 2, they purchase new vehicles in year iny_r. In this case, because they do not need to engage in any additional compliance methods for the VTR in year T_r, the transition in Fig. 3.2 is a-b-e-g-h-i-j.

First, the model for calculating yearly emissions when owners choose Method 1, $E1_{w,r}^{t,p}$, is explained. Here, the residual ratio of PM emission that DPFs cannot eliminate is denoted as rr^p. DPFs do not have any effect on the NOx emission intensity.

Even in Method 1, because owners do not face any restriction before year iny_r, the first equation in Eq. (3.12) is the same as the first one in equation (3.11). In year iny_r, owners install DPFs into their vehicles and use them until the banned year T_r. The second equation in Eq. (3.12) represents the emissions made in this period. The third equation represents emissions after year T_r and is equal to the second equation in Eq. (3.11). Therefore, in Method 1, with the OR, the second equation in Eq. (3.12) is added between the first and second ones in Eq. (3.11).

$$E1_{w,r}^{t,p} = \begin{cases} \left[e_r^p \times N_r^t + e_0^p \times \left(N_r^{2003} - N_r^t\right)\right] \times D & ,if \ \ t < iny_r \\ \left[e_r^p \times rr^p \times N_r^t + e_0^p \times \left(N_r^{2003} - N_r^t\right)\right] \times D & ,if \ \ iny_r \leq t < T_r \\ \left[e_0^p \times NR_r^{T_r} + e_0^p \times \left(N_r^{2003} - N_r^{T_r}\right)\right] \times dr_r^t \times D & ,if \ \ T_r < t \end{cases} \tag{3.12}$$

Second, the explanation for yearly emissions in Method 2 is introduced. In this case, the yearly emissions before year iny_r is also same as the first equations in Eqs. (3.11) and (3.12). The second equation in Eq. (3.11) and the third one in

[10] Let the total amount of mileage be $\overline{TD}_r$, which is constant over time. dr_r^t is described as follows:

$$\begin{aligned} \overline{TD}_r &= D \times dr_r^t \times \left(N_r^{2003} - N_r^t + NR_r^t\right) & ,if \ \ t \neq 2003 \\ &= D \times N_r^{2003} & ,if \ \ t = 2003. \end{aligned}$$

Eq. (3.12) are applied earlier in Eq. (3.13) because old vehicles are replaced with new ones in year iny_r. There is no need to specify the effect of the VTR because any additional actions are not done in the terminal year.

$$E2^{t,p}_{w,r} = \begin{cases} \left[e^p_r \times N^t_r + e^p_0 \times \left(N^{2003}_r - N^t_r\right)\right] \times D & ,if \ \ t < iny_r \\ \left[e^p_r \times NR^{iny_r}_r + e^p_0 \times \left(N^{2003}_r - N^{iny_r}_r\right)\right] \times dr^t_r \times D & ,if \ \ iny_r \leq t \end{cases} \tag{3.13}$$

3.4.3 *Total Benefit of Operational Regulation*

The emission reduction resulting from the OR is the difference between emissions with and without the OR. In Fig. 3.2, □bcfe and □bchg correspond to the concepts of emission reductions in Methods 1 and 2, respectively.

Therefore, with the indicator variable I_r, yearly emission reductions of pollutant p in year t are calculated as follows:

$$TER^{t,p} = \sum \sum_r \left[E^{t,p}_{wo,r} - I_r \times E1^{t,p}_{w,r} - (1 - I_r) \times E2^{t,p}_{w,r}\right]. \tag{3.14}$$

where the first Σ represents summing over vehicle types. When MEC^p is set to be the marginal externality cost (MEC) of pollutant p, the total emission reduction can be converted into monetary value. For simplification, this chapter assumes that the MECs are constant over time. Therefore, the total benefit of the OR evaluated in 2004 is as follows:

$$TB = \sum_t \sum_p \left[TER^{t,p} \times MEC^p \times \exp\{-i \times (t - 2004)\}\right]. \tag{3.15}$$

3.5 Cost-Benefit Analysis

The calculation models for the cost and benefit of the OR require some additional data besides SAP. Table 3.4 indicates all data used in this chapter. Following Oka et al. (2007), it is assumed that no vehicles are in use for longer than 22 years. Therefore, the analytical period is from 2003 (when the OR began) to 2024. Some assumptions must be introduced for calculation purposes: the price of a DPF is estimated to be 400,000 yen, and the discount rate is estimated to be 3 %.[11]

The MECs for NOx and PM are 581 and 22,763,000 yen, as presented by Chap. 2, respectively. Table 3.5 shows the result of the cost-benefit analysis of the OR. The estimation result shows that DPFs are installed on 17,044 vehicles and

[11] The true price is not publicly well known because the market for DPFs is a business-to-business market. Yokemoto and Hiruta (2004) reported that DPFs are sold for between 300,000 and 400,000 yen. Therefore, this assumption is not far from the unknown true price.

Table 3.4 List of variables

Name		Data source
Emission intensity	*e*	Suri-Keikau Inc. (2005)
Passenger vehicle price	*P*	Japan Automobile Dealers Association (2000)
Truck vehicle price	*P*	Japan Trucking Association (2004)
Number of vehicles	*NT*	SAP from AIRIA
Terminal year	*T*	Arimura and Iwata (2008)
Reduced year	*Y*	Arimura and Iwata (2008)
Survival rate	*s*	Trend on Japanese Automobile Possession from AIRIA
Mileage	*D*	MLITT (2001)
Repurchasing rate	*rp*	JAMA (2005a, b)
Depreciation rate	*sr*	Kuroda et al. (1997)
Marginal externality cost	*MEC*	PM: National Environmental Technology Centre (2002)
		NOx: Koyama and Kishimoto (2001)
PM residual ratio	*rr*	8Capital Pref City Aozora Network (2007)
Discount rate	*i*	Assumption (3 %)
DPF price	*Pd*	Assumption (baseline: 40,000 yen)

Table 3.5 Result of cost-benefit analysis by regulation (100 million yen)

	Cost-benefit ratio	Net benefit	Total cost	Total benefit
Operational regulation	7.31	2,279	361	2,640
Vehicle type regulation	2.31	6,812	5,210	12,022

Note: The values of the vehicle type regulation are quoted from Chap. 2

the other vehicles are repurchased with new ones. The total cost of the OR is 36.1 billion yen, whereas the total benefit is 264 billion yen. The net benefit, one of the criteria of whether or not the policy is valid, is positive (227.9 billion yen). Therefore, it is concluded that the OR implemented in the Tokyo region generates enough net benefit to be a valid policy against air pollution.

The bottom of Table 3.5 represents the result of the VTR quoted from Chap. 2. The cost and benefit are 521 and 1,202.2 billion yen, respectively. The difference between them, that is, the net benefit, is 681.1 billion yen.

Figures 3.3 and 3.4 describe the predicted trajectories of PM and NOx emissions only in the Tokyo region. The differences between the "W/O either" and "With VTR" series in the figures are reductions in emission generated by the VTR. Difference between the "With VTR" and "With VTR and OR" series indicate reductions resulting from the OR.

The VTR decreases the total NOx and PM emissions by about 18 and 55 %, respectively, in the Tokyo region (see Figs. 3.3 and 3.4) compared with no regulations. With regard to the OR, it generates additional NOx and PM reductions of 7 and 23 %, respectively, relative to the VTR alone. Therefore, with both regulations, the total NOx and PM emissions fall by 25 and 78 %, respectively, from 2003 to 2024.

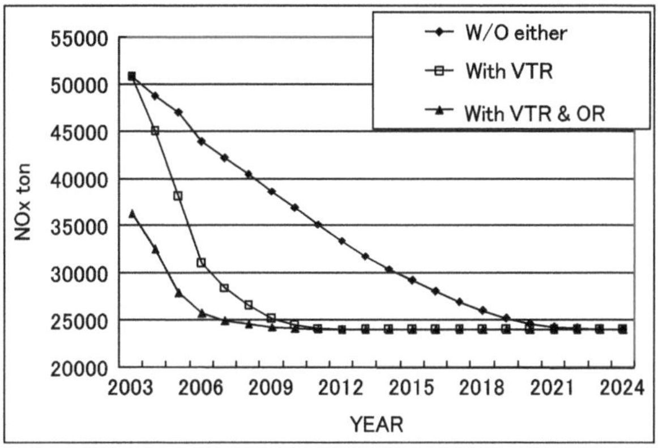

Fig. 3.3 Transition of NOx emissions in Tokyo region (ton)

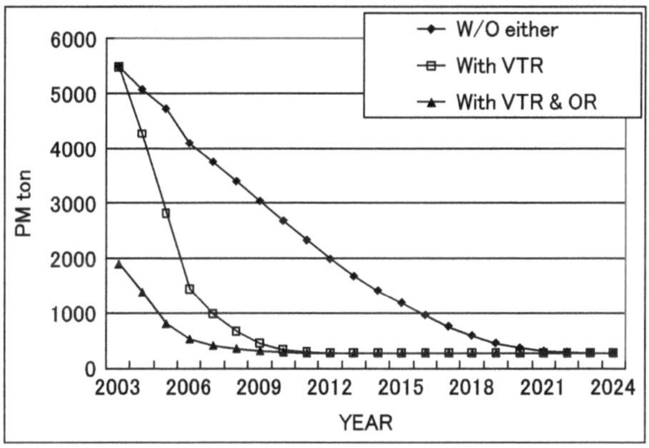

Fig. 3.4 Transition of PM emissions in Tokyo region (ton)

3.6 Robustness Check

The previous result is calculated with the most credible values available, as in the case of the prices of MECs and DPF prices, but they may not represent the precise true value. For a robustness check on the result, this section describes the results of a sensitivity analysis in which two assumptions were changed.

First, the assumptions about MEC were changed. It is well known that there is uncertainty surrounding MEC. For example, although it has been said that PM causes increased incidences of lung cancer and breathing problems, it was recently pointed out that it also has effects on the risks of pollen allergies, apoplexy and cardiac affection. Therefore, the uncertainty of the causal relation between pollutants and health damage is considered to affect the estimated results.

To take this concern into account, we take the approach to similar to Chap. 2. Namely, upper and lower bound MECs were used in the calculation instead of the value used previously. The results are shown in Table 3.6. With the upper bound

Table 3.6 Result of cost-benefit analysis with other MECs (100 million yen)

	MEC	Cost-benefit ratio	Net benefit	Total cost	Total benefit
Operational regulation	Upper bound	10.2	3,319	361	3,680
Vehicle type regulation		3.21	11,540	5,210	16,751
Operational regulation	Lower bound	4.43	1,239	361	1,600
Vehicle type regulation		1.4	2,083	5,210	7,294

Note: The values of the vehicle type regulation are quoted from Chap. 2

MEC value (PM and NOx values of 31,926 and 769 thousand yen per ton), the net benefit of the OR increases to 331.9 billion yen, whereas with the lower bound (PM and NOx values are 13,600 and 394 thousand yen per ton), it decreases to 123.9 billion yen because pollutants are regarded to generate less health damage.[12] It is remarkable that the net benefit remains positive, even in the case of the lower bound MEC.

Next, the DPF price and the selling ratio were changed. It is known that a DPF's price depends on vehicle characteristics: a DPF for large standard trucks is more expensive than that for small trucks. However, appropriate data on DPF pricing could not be obtained. Consequently, a sensitivity analysis on the price was conducted in response to this point. According to cost-minimization behavior, owners are assumed to sell their old vehicles in the second-hand markets of non-regulated areas when repurchasing new ones; that is, the selling ratio is one. However, they may also scrap old vehicles without selling them. The ratio captures this circumstance. Changing the ratio is synonymous with changing the relative prices of transaction prices and old vehicles' residual values. If the ratio is less than one, the transaction prices are less than the residual values because the residual value is invariable and the transaction prices may vary due to asymmetric information between owners and buyers.

Figure 3.5 exhibits the iso-net benefit curve in simultaneously changing both DPF price and the selling ratio. Basically, the upper-right spheres in the figure indicate a large net benefit. Therefore, it is fairly certain that the net benefit is an increasing function of both of them.

This figure shows an important implication for policy making against air pollution. That is to say, a complement policy that increases incentives to move toward the top-right in the figure is recommended. In this analysis, there are only two compliance methods for the OR: purchasing a DPF or purchasing a new clean vehicle. Therefore, an increase in the price of a DPF indicates that the relative cost of purchasing a new vehicles falls. Therefore, a subsidy for new vehicles or a tax on

[12] The upper and lower bound MECs for PM and NOx are presented in Chap. 2.

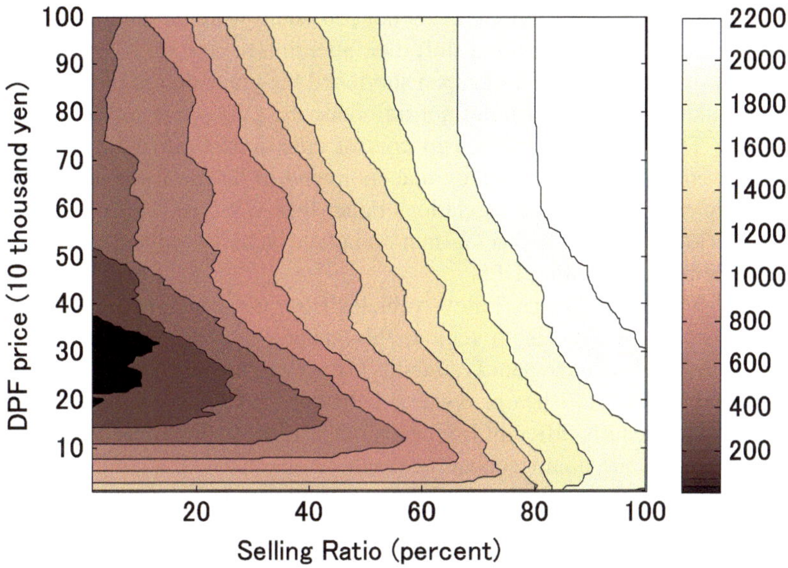

Fig. 3.5 Iso-net benefit curve (100 million yen)

DPFs is adequate as a complementary policy. Conversely, if policy makers introduce a subsidy for DPFs, the improvement of the net benefit would lessen.

This finding mainly comes from the fact that there is a gap in improving PM emission intensities between the installation of DPFs and the purchase of new vehicles. Take as an example a standard truck registered in 1990 weighing 3 tons. Suri-Keikau Inc. (2005) reported that the PM emission intensity of such trucks is 0.154 g/km.[13] DPFs are required to reduce PM emissions from 30 to 70 %. If DPFs reduce the intensity by the maximum 70 %, it drops to 0.046 g/km. On the contrary, the intensity of new standard trucks with the same weight in 2003 is dramatically better, at 0.004 g/km. Therefore, considering the costs and improvements, it is concluded that the replacement of old trucks by new ones is more cost effective than DPF installation. Although the net benefit increases if the DPF price is lower than about 200,000 yen, there is little possibility that this situation would occur as this chapter set the price to be 400,000 yen in Section 3.5.

To further improve social welfare, there are two implications. First, Tokyo-area municipalities should rethink their complementary policies. For example, the Tokyo municipality spent more than 10 billion yen on subsidies for DPF diffusion from 2001 to 2005, whereas their subsidy for new vehicle replacement totaled only 1.4 billion yen. This budget allocation, which can be observed in most

[13] This value reflects the assumptions that vehicles run at a speed of 20 km/h and that there is no deterioration with increasing vehicle age.

municipalities in the Tokyo region,[14] is not consistent with the finding. Therefore, Tokyo-area municipalities should shift their attention to new vehicle subsidies.

Second, if the government of Japan shortened the grace period of the VTR and the Tokyo-area municipalities did not introduce the OR, social welfare would be increased. There are few policies to control emissions from old dirty existing vehicles in the world. When policy makers intend to handle these vehicles, they should implement VTR-type regulations rather than OR-type regulations because the cost effectiveness of replacing non-compliant vehicles with new vehicles is better than that of installing DPF.

Although the OR restricts inflow vehicles from outside Tokyo in addition to those registered in the Tokyo region, this chapter only investigates the latter category of vehicles due to data availability. However, the result is not undermined unless vehicle characteristics or owners' behaviors inside and outside the Tokyo region are significantly different from each other. If the vehicles inside and outside the Tokyo region are homogenous, the result can be updated, adding the cost and benefit of inflow vehicles. MLITT (2004) reported that 18 % of the cargo volume for the Tokyo municipality came from other municipalities. Therefore, by dividing the cost and benefit in Table 3.5 by 0.82, the updated values can be obtained. The calculated net benefit is 277.9 billion yen.

Because this chapter uses much of same data as Chap. 2, it has basically the same biases found there. The cost calculation model may be overestimated because maintenance and operation costs, which are lower in new vehicles than in old ones, are ignored due to a lack of available data. In addition, the lack of distinction between purchasing new vehicles and old used vehicles that comply with the OR also leads to the overestimation of costs.

On the other hand, the benefit calculated in this chapter may be underestimated because this chapter assumes no innovation on emission intensity and treats MEC only for health damages. However, the overestimation and underestimation do not definitely invert the result that the OR is valid as an air pollution countermeasure. These ignored components should be incorporated into future evaluations of air pollution policy.

3.7 Conclusion

This chapter examines the cost-benefit analysis of the OR, uniquely implemented in the Tokyo region municipalities to reduce air pollution from automobiles. With the cost and benefit calculation models, it was revealed that the benefit generated from the regulation is larger than the cost. Therefore, it is concluded that the OR is valid as an air pollution policy.

---

[14] Only the municipality of Kawasaki City spent more in subsidies on new vehicle replacement than on DPFs in the same period.

However, this result does not ensure that the OR is efficient from the perspective of economic theory. The sensitivity analysis shows that it is more cost effective to replace dirty old vehicles with clean new ones than it is to install DPFs into old ones. Therefore, municipalities should adopt subsidies for promoting vehicle replacement as a complementary policy for the OR.

In Japan, all vehicles are given identification numbers, like matriculation numbers, through the legal automobile registration and inspection system. Making this system information publicly available will make it possible to perform more accurate policy evaluations and thus lead to more efficient policy planning.

References

8Capital Pref City Aozora Network (2007) Guideline on designating diesel particulate filter in 8Cpalital Pref City (in Japanese). http://www.8taiki.jp/regulatory/pdf/youkou.pdf. Accessed in 25 July 2009

Arimura TH, Iwata K (2008) Economics analysis on motor-vehicle type regulation: policy evaluation of NOx-PM law (in Japanese). Environ Sci 21(2):103–114

Austin D, Dinan T (2005) Clearing the air: the costs and consequences of higher CAFE standards and increased gasoline taxes. J Environ Econ Manage 50(3):562–582

Dill J (2004) Estimating emissions reductions from accelerated vehicle retirement programs. Transport Res Part D 9(2):87–106

Goldberg PK (1998) The effects of the corporate average fuel efficiency standards in the US. J Ind Econ 46(1):1–33

Hibiki A, Arimura T H (2005) An empirical study of the effect of the fuel tax in Japan on vehicle selection and NOx emission. Department of Social Engineering Discussion Papers at Tokyo Institute of Technology NO.05-02

Iwata K, Arimura TH (2009) Economic analysis of Japanese air pollution regulation: an optimal retirement program under the vehicle type regulation in the NOx-particulate matter law. Transport Res Part D 14(3):157–167

Japan Automobile Dealers Association (JADA) (2000) Annual report on new vehicle registration 23 (in Japanese). JADA, Tokyo

Japan Automobile Manufacturers Association (JAMA) (2005a) Standard truck market survey 2004 (in Japanese). JAMA, Tokyo

JAMA (2005b) Small truck market survey 2004 (in Japanese). JAMA, Tokyo

Japan Trucking Association (JTA) (2004) Japanese truck freight transportation 2004 (in Japanese). JTA, Tokyo

Kleit AN (2004) Impacts of long-range increases in the fuel economy (CAFE) standard. Econ Inq 42(2):279–294

Koyama S, Kishimoto A (2001) External costs of road transport in Japan (in Japanese). Transport Policy Stud Rev 4(2):19–30

Kuroda M, Shinpo K, Nomura K, Kobayashi N (1997) KEIO database: measurement of output, labor and capital inputs (in Japanese), vol 8, Keio economic observatory monograph series. Keio University, Tokyo

Lopez JM, Jimenez F, Aparicio F, Flores N (2009) On-road emissions from urban buses with SCR plus Urea and EGR plus DPF systems using diesel and biodiesel. Transport Res Part D 14 (1):1–5

Lumbreras J, Valdes M, Borge R, Rodriguez ME (2008) Assessment of vehicle emissions projections in Madrid (Spain) from 2004 to 2012 considering several control strategies. Transport Res Part A 42(4):646–658

Ministry of Land, Infrastructure, Transport and Tourism (MLITT) (2001) Monthly statistical report on motor vehicle transportation 2001 March (in Japanese). MLITT, Tokyo

MLITT (2004) Land and transport statistics directory (in Japanese). MLITT, Tokyo

National Environmental Technology Centre (2002) Estimates of the marginal external costs of air pollution in Europe. http://ec.europa.eu/environment/enveco/air/pdf/betaec02a.pdf. Accessed in 25 July 2009

Oka T, Fujii Y, Ishikawa M, Matsuo Y, Susami S (2007) Maximum abatement costs for calculating cost-effectiveness of green activities with multiple environmental effects. In: Huppes G, Ishikawa M (eds) Quantified eco-efficiency (eco-efficiency in industry and science). Springer, Dordrecht, pp 41–78

Sano M (2008) Genealogy of automotive technology innovation and technology policy in Japan (in Japanese). In: Miyoshi H, Tanishita M (eds) Technological innovation in the automotive industry and economic welfare. Hakuto Shobo, Tokyo, pp 1–36

Suri-Keikau Inc. (2005) Vehicle emission intensities and total emission estimation survey 2004: commission research report to the ministry of the environment (in Japanese). Suri-Keikau Inc., Tokyo

Yokemoto M, Hiruta K (2004) Motor vehicle pollution control and cost allocation: focus on motor vehicle NOx/PM law and diesel access restrictions in Tokyo (in Japanese). Res Environ Disrupt 33(4):25–32

Chapter 4
Does Environmental Regulation Affect on Outside of the Regulated Areas? Empirical Analysis of Japanese Automobile NOx-PM Act

Abstract The Automobile NOx-PM Act was introduced to mitigate air pollution problem in the Japanese metropolitan areas in 2001. Many old fume-spewing vehicles might flow outside the areas because the regulation prohibited their usage in the regulated areas only. To compare secondary vehicle market prices before and after the implementation, this chapter examines whether the price changed by the outflow due to the regulation. The estimation results show that the regulation did not lower the prices statistically significantly. In contrast, we found increase used tracks' export. These facts indicate that diffusion of low-emission vehicles in foreign countries might be discouraged by reinforcement of Japanese environmental regulation, which might be a type of pollution haven hypothesis. More attention on secondary market needs be paid when regulators intend to introduce environmental regulation on goods such as automobiles.

Keywords Automobile NOx-PM Act • Air Pollution • Ancillary impact • Hedonic Approach • Used Car Markets

4.1 Introduction

Various environmental programs have been utilized as policy instruments to control vehicle emissions. For example, emission standards in the U.S., Europe, and Japan place upper limits on emissions intensity, and the Japanese Top Runner Program requires vehicles to meet the program's energy efficiency standards. These programs typically target only new vehicles since they require manufacturers to produce and sell vehicles that comply with certain environmental standards. It is older vehicles, however, that are generally more polluting and harmful to the environment. The emissions control technology becomes obsolete as vehicles age (National Cooperative Highway Research Program 1997) and yet older vehicles evade most regulations. In order to control emissions from old, fume-spewing vehicles currently under use, authorities initiated programs that promote the replacement of old vehicles.

T.H. Arimura, K. Iwata, *An Evaluation of Japanese Environmental Regulations*,
DOI 10.1007/978-94-017-9947-8_4

In Japan, the Vehicle Type Regulation (hereafter referred to as "VTR") was implemented in 2001 under the NOx-PM Act[1] in three metropolitan areas, namely, Tokyo, Osaka and Nagoya, mainly to reduce nitrogen oxides (NO_x) and particulate matter (PM). VTR specifies terminal years for old vehicles registered in these areas. By prohibiting the use of high-emission vehicles built before a certain point in time, VTR promotes earlier replacement of these vehicles with newer, low-emission ones (Ministry of the Environment 2005).

Chapter 2 empirically examined the cost and benefit of VTR and confirmed that the regulation is cost-effective. We found that VTR substantially contributed to NO_x and PM reduction, resulting in health benefits of 1,202.2 billion yen. It was revealed however, that the time for retiring old vehicles is not optimally scheduled by the current VTR and that the benefit should increase if the timing is effectively altered.

Vehicle retirement programs in other countries have also been found to improve air quality and social welfare. In the U.S., the Delaware Vehicle Retirement Program, a pilot program run between 1992 and 1993, offered an owner of pre-1980 vehicles a $500 subsidy to scrap his or her vehicle for the purpose of reducing hydrocarbons (HC). Alberini et al. (1996) argue that programs of this sort have small environmental effects but they are cost-effective. Dill (2004) examined the vehicle retirement programs in California and found their positive effect on reducing emissions. Lin et al. (2008) analyzed the effectiveness of a fleet scrappage program in Illinois and confirmed its short-term emission benefits. Lumbreras et al. (2008) simulated several scenarios to assess future measures to reduce air pollution in Madrid and found that renewing vehicles is the most effective measure for emissions reductions.

In the U.S., vehicles are usually scrapped at the time of retirement. Little is known, however, as to where vehicles end up after being forced to retire by the Japanese VTR program.[2] One possible scenario is that they are transported and sold in secondary vehicle markets outside the regulated areas. In 2005, a committee of the central council in the Ministry of the Environment addressed this possibility and expressed their concern that the outflow of banned old vehicles from the regulated areas is likely to discourage the diffusion of low-emission vehicles in other areas.

Another possibility is that the vehicles are exported overseas, particularly to developing countries and newly industrialized countries nearby (e.g. Russia) where emissions regulations are not as strict as in Japan. This exemplifies the pollution haven hypothesis. That is, the regulation encourages exporting old vehicles and thereby slows down the increase of new low-emission vehicles in the importing countries. This concern is already raised in Fuse and Kashima (2007a) with regard to the NOx Act, the

[1] The Act is officially named "the Act concerning special measures for the reductions of nitrogen oxides and particulate matter from automobiles in specified areas".

[2] Due to data limitation, we cannot determine whether vehicles that were brought outside of the regulated areas have been regulated by VTR. VTR regulates vehicles based on their emission codes and hence, even among vehicles of the same type and year, some are subject to VTR while others are not.

forerunner of VTR. According to the authors, the NO_x Act indirectly increased the export of old polluting vehicles from Japan to developing countries.

The present chapter explores these two possibilities by empirically examining the effect of VTR on unregulated areas. Specifically, we will investigate whether VTR encouraged the outflow of old vehicles and then influenced secondary vehicle prices in unregulated areas. We do this by comparing secondary vehicle prices before and after the regulation's implementation. If changes in the prices are observed, it confirms the indirect effect of VTR in unregulated areas. If not, it may suggest that retired vehicles have been exported overseas.[3]

Previous studies on VTR evaluated the regulation in terms of its impact on the regulated areas (e.g., Iwata and Arimura 2009). To what extent unregulated areas have been affected has yet to be investigated. This chapter is one of the first to examine the effect of VTR on secondary vehicle markets outside the regulated areas. It is hoped that the findings of this chapter contribute to the ongoing effort for more extensive and comprehensive policy evaluations as described by the Ministry of Internal Affairs and Communications (2007): their analysis aims to "quantitatively analyze the effectiveness of a regulation and identify each and every factor that constitutes the impact of the regulation."

The rest of the chapter is organized as follows: The next section provides an overview of VTR and Section 4.3 introduces estimation models. Section 4.4 describes the dataset and Section 4.5 presents estimation results. Section 4.6 concludes.

4.2 The Effect of VTR on Unregulated Areas

VTR replaced the NOx Act[4] in 2001 with the aim of further reducing vehicle emissions (Sano 2008).[5] With this change, the regulated areas were extended to 276 municipalities, including those in Nagoya, from the original 196 municipalities in Tokyo and Osaka.

---

[3] The Ministry of Economy, Trade and Industry (2006) estimated that 0.1 % of vehicles are illegally dumped. This chapter assumes that of all vehicles regulated by VTR, only a small portion, a percentage similar to the Ministry's estimation, was illegally abandoned. Most regulated vehicles are diesel trucks owned by firms/facilities, who receive more severe penalties than individuals for illegally abandoning their vehicles. Thus, illegal dumping does not appear to be an optimal solution for owners. It is also possible that some of the regulated vehicles are scrapped. However, as will be mentioned later, a relatively small number of vehicles are scrapped.

[4] The regulation was introduced in 1992 and its official name was "the Act Concerning Special Measures for Total Emission Reduction of Nitrogen Oxides from Automobiles in Specified Areas".

[5] VTR was partially revised in 2008 to incorporate additional regulations on vehicles brought over from unregulated areas as well as pollution control measures in target regions. See Ministry of the Environment (2007) for details. These revisions are excluded from our analysis because our focus is the effect of VTR before the revisions were made.

VTR is unique in that it promotes the early retirement of old vehicles by placing bans on them in accordance with their type (e.g., standard truck, standard bus) and initial registration year. For example, a standard truck initially registered in 1990 is prohibited for use as of 2005 and a small truck initially registered in 2000 is prohibited after 2009 (see Chap. 2). That is, even if owners intend to use their vehicles as long as they wish, they must discard their vehicles at the time designated by VTR. Because VTR is mainly concerned with diesel trucks, it exempts low-emission vehicles such as gasoline passenger vehicles and hybrid cars (Ministry of the Environment 2005).

As presented in Chap. 2, approximately 3.9 million vehicles were subject to VTR at the end of March 2003 (excluding gasoline passenger vehicles), among which 2.6 million (66 %) were regulated for being highly polluting. That is, quite a large number of vehicles were regulated that year; it almost exceeds the number of standard trucks initially registered in the country in 2010.

The Japan Automobile Manufacturers Association (JAMA) conducted surveys (2005a, b) in which owners of vehicles subject to VTR were asked which complying methods they adopted. It was found that 89 % of the owners of standard trucks and 83 % of the owners of small or light-duty trucks chose "replacing their vehicles with new complying vehicles." This means that most of the 2.6 million vehicles were replaced upon facing the ban. Because they were forced to retire these vehicles while they still retained some value, it is plausible that the owners considered selling them as used cars to lessen the financial burden of replacement. Quite likely, therefore, many of the retired vehicles were transported and sold in secondary vehicle markets outside the regulated areas.

Figure 4.1 shows the numbers of new and used standard trucks sold and registered in the country from 1998 to 2010.[6] The registration steadily declines from 1998. This is possibly due to the recession and also due to the revisions made to the Energy Saving Act in 2006. The revised act requires private carriers with carrying capacity over a certain standard to formulate energy saving plans and periodically report the amount of energy use. These additional requirements may have discouraged registration.

The number of new trucks sold sharply increases from 2003. This is probably because the first VTR ban became effective in that year, a certain number of vehicles having retired for the first time since the implementation. Accordingly, the number of used trucks sold also increases from 2003, though the increase is not as dramatic as the number of new trucks sold.

Figure 4.2 shows the numbers of used standard trucks initially registered and those scrapped or permanently deregistered in the regulated areas (left) and other areas (right). The number of trucks permanently deregistered in the regulated areas rapidly increased to 20,000 in 2003, presumably because many vehicles complied

[6] The number sold is the number of vehicles sold in that year. The number registered is the number of existing vehicles in that year. In other words, the numbers sold and registered are flow and stock, respectively.

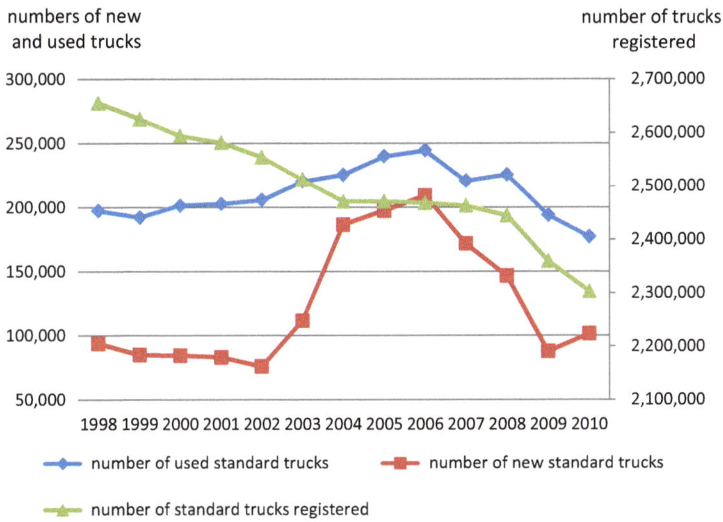

Fig. 4.1 Number of standard trucks sold and registered

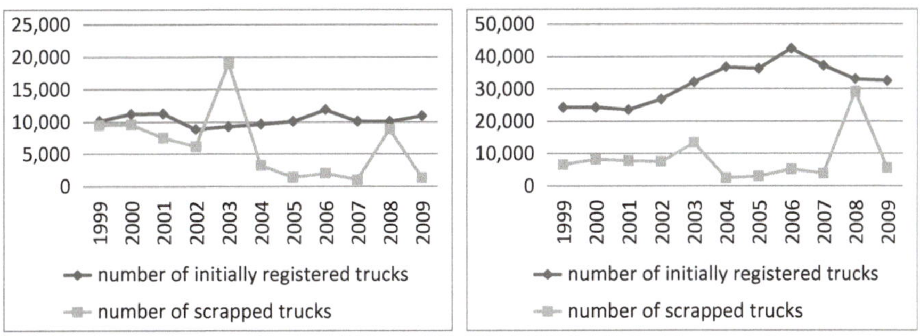

Fig. 4.2 Numbers of used standard trucks initially registered and permanently deregistered in the regulated areas (*left*) and other areas (*right*)

with the first VTR ban.[7] This number is not substantially large, however, given that as many as 580,000 trucks were subject to the regulation at the end of 2003 (see Chap. 2). Then a question arises: What happened to the retired trucks that were not scrapped or permanently deregistered? Some might have been sold outside the regulated areas. In fact, as shown in the figure, the number of initial registrations steadily increases from 2001 in unregulated areas. This is in contrast to the regulated areas where the number has been stable.

[7] The number rises again in 2008. This is presumably because VTR was partially revised in that year. We do not discuss this increase because our analysis is concerned with the impact of VTR before the revisions were made.

If many retired vehicles were indeed transported and sold in secondary vehicle markets outside the regulated areas, the prices of secondary vehicles in those areas should decrease as a result of the increase in supply. In other words, declines in secondary vehicle prices suggest that VTR has an influence on unregulated areas. In the next section, we will discuss this further along with our estimation models.

4.3 Estimation Models

To estimate the effect of VTR on secondary vehicle prices in unregulated areas, we constructed an econometric model based on the hedonic approach. The approach was first theoretically developed by Rosen (1974) and has been used to estimate the implicit prices associated with the attributes of differentiated goods. In the context of vehicles, Van Daren and Bode (2004), Reis and Silva (2006) and Matas and Raymond (2009) used the hedonic approach to construct quality-adjusted price indexes. The approach is also used when researchers attempt to estimate the implicit prices associated with non-marketable goods such as road safety, vehicle pollution and vehicle congestion.

The hedonic approach is suitable for our purpose for two reasons. Secondary vehicles are considered differentiated goods because even if they are the same model, they are not identical in that their mileages and years are individually different. In addition, this approach allows us to consider (1) whether the vehicle was sold after VTR has been implemented and (2) whether the vehicle was sold outside the regulated areas with the attributes of a secondary vehicle.

The hedonic approach requires the market under study to be competitive. This requirement appears to be satisfied by the market we examined. First, there are many firms in the market; as of June 2010 as many as 9,948 firms are registered in the Japan Used Car Dealers Association that aims for fair trade and distributions of secondary vehicles.[8] Second, the market has low barriers to entry; sellers only need to submit the secondhand dealer permit application to the local police department when starting the business.

In what follows, we explain our estimation model and method. We assume that the price of a secondary vehicle i (P_i) depends on a dummy variable that takes one if the vehicle was sold outside the regulated areas (Non_i), a dummy variable that takes one if the vehicle was sold after VTR has been implemented (T_i), the interaction of these variables ($Non_i \cdot T_i$), the emission volume (X_i), and a vector of the vehicle's other attributes such as model, year and mileage ($\mathbf{Z}_i$):

$$P_i = f(Non_i, T_i, Non_i T_i, X_i, \mathbf{Z}_i). \tag{4.1}$$

Equation (4.1) is the hedonic pricing function for secondary vehicles, which represents a lower envelope of a family of the lowest prices for a vehicle that

[8] http://www.jucda.or.jp/.

Table 4.1 Parameters that capture the effect of VTR

	Regulated areas	Other areas
Before the implementation	0	$\alpha 1$
After the implementation	$\alpha 2$	$\alpha 2 + \alpha 3$

individual sellers can offer to achieve a certain level of profit (i.e., sellers' offer functions), under the assumption that they maximize their profit given their technological constraints.

As the appropriate functional form for Eq. (4.1) cannot be specified on theoretical grounds, we embed Eq. (4.1) in the flexible functional form given by the Box-Cox transformation. In particular, P_i and X_i are transformed as $P_i^{(\theta)} = (P_i^\theta - 1)/\theta$ and $X_i^{(\lambda)} = (X_i^\lambda - 1)/\lambda$, respectively, where θ and λ are parameters, while the other variables (i.e., Non_i, T_i and $\mathbf{Z}_i$) are not transformed. This is because the Box-Cox transformation can be applied to variables that only take positive values; P_i and X_i are always positive, while Non_i, T_i and $\mathbf{Z}_i$ are not. As a result, the estimating equation becomes

$$P_i^{(\theta)} = \alpha_0 + \alpha_1 Non_i + \alpha_2 T_i + \alpha_3 Non_i \cdot T_i + \alpha_4 X_i^{(\lambda)} + \mathbf{Z}_i \boldsymbol{\beta} + \varepsilon_i \tag{4.2}$$

where ε_i is an error term. Equation (4.2) takes various forms depending on the values of θ and λ. Specifically, Eq. (4.2) allows us to have a level-level specification, a level-log specification, a log-level specification and a log-log specification for P_i and X_i. The estimation of θ and λ is essentially equivalent to choosing the functional form that best fits the data.

In this chapter, we use two periods, 2001 and 2007, for estimation. The former corresponds to the period before and the latter to after the implementation of VTR. The effects of VTR are summarized in Table 4.1 by areas (i.e., regulated or not regulated) and across time (i.e., before and after the implementation). α_3, i.e., the coefficient on $Non_i \cdot T_i$, is the parameter of interest, representing the effect of VTR on secondary vehicle prices outside the regulated areas. If α_3 is negative, then it means that VTR resulted in lowering secondary vehicle prices outside the regulated areas.

Given Eq. (4.2), the concentrated log-likelihood function, $\ln L^*$, is expressed as

$$\ln L^*(\theta, \lambda) = \left(-\frac{N}{2}\right)\left\{\ln(2\pi) + \ln\left[\hat{\sigma}^2(\theta, \lambda)\right] + 1\right\} + (\theta - 1)\sum_{i=1}^{N} \ln(P_i) \tag{4.3}$$

where N is the sample size and $\hat{\sigma}^2$ is an estimator of the variance of the error term:

$$\hat{\sigma}^2 = \frac{1}{N}\sum_{i=1}^{N}\left\{P_i^{(\theta)} - \hat{\alpha}_0 - \hat{\alpha}_1 Non_i - \hat{\alpha}_2 T_i - \hat{\alpha}_3 Non_i \cdot T_i - \hat{\alpha}_4 X_i^{(\lambda)} - \mathbf{Z}_i \boldsymbol{\beta}\right\}^2. \tag{4.4}$$

Let $\boldsymbol{\delta} = [\alpha_0, \ldots, \alpha_4, \boldsymbol{\beta}^T]$ and $\mathbf{W} = \left[\mathbf{1} \quad \mathbf{Non\ T} \quad \mathbf{Non \cdot T\, X}^{(\lambda)} \quad \mathbf{Z}\right]$. Then, an estimator of $\boldsymbol{\delta}$, $\widehat{\boldsymbol{\delta}}$, is given as

$$\widehat{\boldsymbol{\delta}} = \left(\mathbf{W}^T\mathbf{W}\right)^{-1}\mathbf{W}^T\mathbf{P}^{(\theta)} \tag{4.5}$$

4.4 Data and Explanatory Variables

The key variables for identifying the effect of VTR on secondary vehicle markets are vehicle prices in secondary markets before and after the implementation of VTR and the attributes of those vehicles. We used "Car Sensor," one of the most popular used-car magazines for our database. Because VTR mainly targets trucks, our analysis focuses on five leading models of top-selling truck manufacturers, namely, Elf (Isuzu), Canter (Mitsubishi Fuso), Toyoace (Toyota), Dyna (Toyota) and Atlas (Nissan).

Table 4.2 shows the number of trucks sold in 2007 in descending order based on a truck information site, "Truck-next." As shown in the table, all five models under study are ranked among the six most sold trucks, holding 80 % of the market share: Dyna and Toyoace ranked first (sixth in the table where vans are also ranked among trucks), followed by Elf, then Canter, Titan and, finally, Atlas. Hence, we believe that the amount of bias due to limiting our sample to the five trucks is rather small.

Table 4.2 Number of light trucks and vans sold in 2007

Rank	Manufacture name	Vehicle name	Vehicle types	Number of vehicles sold
1	Toyota	Hi-ace and Regius ace	Van	63,305
2	Toyota	Probox van	Van	41,588
3	Nissan	AD van	Van	38,563
4	Nissan	Vanette	Van	17,153
5	Toyota	Succeed	Van	16,814
6	Toyota	Dyna and Toyoace	Truck	16,242
7	Mazda	Bongo	Van	14,989
8	Nissan	Caravan	Van	12,208
9	Isuzu	Elf	Truck	12,150
10	Mitsubishi Fuso	Canter	Truck	11,631
11	Toyota	Town ace	Van	8,637
12	Toyota	Lite ace	Van	7,002
13	Mazda	Titan	Truck	6,143
14	Nissan	Atlas	Truck	5,828
15	Hino	Dutro	Truck	5,192
16	Mazda	Familia	Van	3,824
17	Mitsubishi	Lancer	Van	3,350
18	Honda	Partner	Van	2,782
19	Mitsubishi	Delica	Van	2,454
20	Mazda	Brawny	Van	2,302
21	Isuzu	Como	Van	597
22	Nissan	Expert	Van	156
23	Nissan Diesel	Condor	Truck	104

Table 4.3 Sample sizes in the regulated/other areas and before/after the implementation

	Regulated areas	Other areas	Subtotal	
Before the implementation	197	139		336
After the implementation	89	987		1,076
Subtotal	286	1,126	Total	1,412

Note that Titan, ranked fifth (thirteenth in the table), is excluded from our analysis because the model is an original equipment manufacturer.[9]

With regard to the period after the implementation of VTR, we obtained the prices and attributes of the five models from "Car Sensor Net," the online version of the magazine (retrieved at the end of September 2007). Because we were not able to access Car Sensor Net for the period before the implementation, we instead used hard copies of the magazine. Every issue of the magazine comprises five editions, each covering one of the following regions: Kanto, Tokai, Kansai, Hokkaido and Kyushu. We used the issue released at the beginning of every month from November 1999 to April 2000; that is, five editions of the magazine for the period of 6 months and, therefore, a total of 30 issues.

When a truck is included in multiple issues, its price is likely to be different in each issue. To avoid having the identical vehicle in our dataset, we only used the latest issue when a truck was listed multiple times.

Table 4.3 shows our sample sizes by area (i.e., regulated areas and other areas) and by period (i.e., before and after the implementation). It is 286 and 1,126 in the regulated areas and other areas, respectively, and 336 and 1,076 before and after the implementation, respectively. We believe that our sample is sufficiently large to identify the effect. It should be noted that our data are not panel. That is, each and every vehicle under study (i.e., vehicles in 2000 and those in 2007) is different and there are no identical vehicles in our dataset.[10]

Below we explain the variables used in this chapter. The dependent variable P in Eq. (4.1) is adjusted by the consumer price index for vehicles in 2005, given the difference in the price levels between 2000 and 2007. Emission volume (*EMVOLUME*), expressed in liter, indicates the engine size. Vehicle age (*AGE*) is the number of years passed since the vehicle was initially registered. *MILEAGE* is the total mileage driven by the vehicle and it is expressed in the unit of 10,000 km. *INSPECTION* is a variable that represents the number of months remaining until the next inspection.[11] It is mandatory to have one's vehicle inspected every certain number of years, and the variable takes zero if the inspection has expired. To capture a vehicle's repair history, we use a dummy variable *REPAIR* that takes one if the vehicle has been repaired. *SINGLE* takes one if a vehicle is a single owner car.

[9] The supplier of Titan is Isuzu Elf.

[10] There are some studies that analyze identical vehicles sold repeatedly. This approach is known as the repeat sales method (see Meese and Wallace 1997).

[11] As for the automobile inspection system in Japan, refer to Chap. 3.

We also have dummy variables for four-wheel drive (*FWDRIVE*), antilock brake system (*ABS*), air-conditioning (*A/C*), automatic transmission (*AT*), driver frontal air bag (*D-AIRBAG*), passenger frontal air bag (*P-AIRBAG*), keyless entry (*KEYLESS*), rear view monitor (*MONITOR*) power windows (*PWINDOWS*), power steering (*PSTEER*), recycling fee (*RECYFEE*), periodic inspections (*PINSPC*), complying with VTR (*COMPLYING*), PM removal equipment (*PM*), and unregulated areas (*NON*). These variables take one if a vehicle is equipped with the corresponding device/function or if it meets the corresponding condition. An exception is *RECYFEE*, which represents whether the vehicle's recycling fee has been paid in accordance with the Automobile Recycling Act.[12] Because the Act was implemented in 2002, this dummy variable is zero for all vehicles in 2000.

The following specification dummies are also included because there are many types of trucks to suit various purposes: *RV* (recreational vehicle), *REEFER* (reefer, or refrigerated/freezer truck), *INSULATED* (insulated truck[13]), *BUCKET* (bucket truck), *CRANE* (crane truck), *DUMP* (dump truck), and *TAILGATE* (truck with a tailgate lift).

Table 4.4 shows the descriptive statistics of our sample in 2000 and 2007. As shown in the table, the average price of secondary vehicles increased to 1.44 million yen after the implementation of VTR from 1.18 million yen before the implementation. The standard deviation is larger after the implementation as well. Compared to those in 2000, vehicles are better equipped in 2007 and thus we cannot simply conclude that vehicle prices increased after the implementation. When the 2 years are compared, some variables are significantly different from each other (e.g., *EMVOLUME*, *INSPECTION*), especially vehicle equipment variables such as *ABS*, *D-AIRBAG*, and *A/C*.

4.5 Estimation Results

Table 4.5 shows our estimation results. Columns 1 and 2 present the results of model 1 where Box-Cox transformation is applied only to the dependent variable, i.e., the vehicle price, and the results of model 2 where the transformation is applied to emission volume as well as the vehicle price, respectively.

To examine whether our Box-Cox specification outperforms restrictive models, we conducted likelihood ratio tests. For model 1, we examined three particular restrictions, (1) $\theta = 1$, (2) $\theta = 0$ and (3) $\theta = -1$, which correspond to no transformation (i.e., $P_i - 1$), log transformation (i.e., $\ln P_i$) and reciprocal transformation (i.e., $1 - 1/P_i$), respectively. As presented in the last row of Column (1), the test results show that all of these restrictions are rejected. Similar results are obtained

[12] The act is officially named "the Act for the recycling of end-of-life vehicles."

[13] It is a truck with a container whose wall is filled with insulating materials for the purpose of keeping the cargo cold.

Table 4.4 Descriptive statistics by year

Variables	2000 Mean	s.d.	Min	Max	2007 Mean	s.d.	Min	Max	t-test
PRICE	118.38	68.36	17.8	468.9	144.29	79.81	19.0	652.4	***
EMVOLUME	3.52	0.78	1.5	5.3	4.03	0.80	1.6	6.3	***
AGE	5.70	3.38	0.0	25.0	8.82	3.52	0.0	24.0	***
MILEAGE	5.49	4.43	0.0	25.6	8.11	5.18	0.0	29.8	***
INSPECTION	6.76	5.78	0.0	24.0	3.03	4.99	0.0	24.0	***
REPAIR	0.01	0.09	0.0	1.0	0.02	0.12	0.0	1.0	
SINGLE	0.17	0.38	0.0	1.0	0.08	0.26	0.0	1.0	***
FWDRIVE	0.07	0.25	0.0	1.0	0.08	0.27	0.0	1.0	
ABS	0.02	0.13	0.0	1.0	0.10	0.30	0.0	1.0	***
A/C	0.83	0.38	0.0	1.0	0.96	0.19	0.0	1.0	***
AT	0.07	0.26	0.0	1.0	0.09	0.28	0.0	1.0	
D-AIRBAG	0.02	0.13	0.0	1.0	0.15	0.35	0.0	1.0	***
P-AIRBAG	0.00	0.00	0.0	0.0	0.00	0.07	0.0	1.0	
KEYLESS	0.00	0.00	0.0	0.0	0.02	0.12	0.0	1.0	**
MONITOR	0.01	0.08	0.0	1.0	0.02	0.15	0.0	1.0	**
PWINDOWS	0.52	0.50	0.0	1.0	0.84	0.36	0.0	1.0	***
PSTEER	0.95	0.21	0.0	1.0	0.97	0.16	0.0	1.0	**
RECYFEE	0.00	0.00	0.0	0.0	0.66	0.47	0.0	1.0	***
PINSPC	0.17	0.38	0.0	1.0	0.14	0.35	0.0	1.0	
COMPLYING	0.00	0.00	0.0	0.0	0.17	0.38	0.0	1.0	***
PM	0.00	0.00	0.0	0.0	0.01	0.07	0.0	1.0	*
RV	0.03	0.16	0.0	1.0	0.01	0.10	0.0	1.0	***
REEFER	0.03	0.16	0.0	1.0	0.04	0.20	0.0	1.0	*
INSULATED	0.00	0.05	0.0	1.0	0.01	0.12	0.0	1.0	**
BUCKET	0.01	0.09	0.0	1.0	0.03	0.17	0.0	1.0	**
CRANE	0.04	0.21	0.0	1.0	0.12	0.32	0.0	1.0	***
DUMP	0.01	0.12	0.0	1.0	0.04	0.19	0.0	1.0	**
TAILGATE	0.08	0.27	0.0	1.0	0.07	0.25	0.0	1.0	
NON	0.41	0.49	0.0	1.0	0.92	0.28	0.0	1.0	***

Note: the units of PRICE, EMVOLUME, MILEAGE and INSPECTION are 10,000 yen, litter, 10,000 km and month, respectively

for model 2 when we examine the following restrictions, (1) $\theta = 1$, $\lambda = 1$ (2) $\theta = 0$, $\lambda = 0$ and (3) $\theta = -1$, $\lambda = -1$. These results imply that our Box-Cox specification outperforms conventional specifications such as the linear or the log-linear specification.

The coefficient on the interaction term, $Non \cdot T$, is found to be insignificant both in models 1 and 2. This means that regardless of whether we Box-Cox transform the dependent variable or both the dependent variable and emission volume, VTR did not affect secondary vehicle prices in unregulated areas.

Our results also indicate that consistent with our intuition, the price of a vehicle increases if it is newer, its emission volume is larger, and/or its mileage is lower.

Table 4.5 Estimation results

	Model 1		Model 2	
Independent variables	Coef.	P-value	Coef.	P-value
EMVOLUME	0.416	0.00	0.398	0.00
AGE	−0.127	0.00	−0.127	0.00
MILEAGE	−0.032	0.00	−0.032	0.00
INSPECTION	0.005	0.10	0.005	0.10
REPAIR	0.061	0.59	0.061	0.59
SINGLE	−0.070	0.15	−0.071	0.15
FWDRIVE	0.399	0.00	0.401	0.00
ABS	0.011	0.85	0.011	0.86
A/C	−0.090	0.15	−0.090	0.15
AT	−0.109	0.04	−0.109	0.04
D-AIRBAG	0.099	0.07	0.099	0.07
P-AIRBAG	0.002	0.99	0.004	0.99
KEYLESS	0.156	0.20	0.156	0.20
MONITOR	0.001	0.99	0.000	1.00
PWINDOWS	0.200	0.00	0.201	0.00
PSTEER	−0.040	0.63	−0.040	0.63
RECYFEE	−0.039	0.24	−0.040	0.24
PINSPC	−0.096	0.02	−0.096	0.02
COMPLYING	0.071	0.61	0.072	0.61
COMPLYING×NON	0.057	0.70	0.056	0.71
PM	0.301	0.14	0.300	0.15
RV	1.961	0.00	1.967	0.00
REEFER	0.580	0.00	0.581	0.00
INSULATED	0.146	0.24	0.146	0.24
BUCKET	1.092	0.00	1.096	0.00
CRANE	0.901	0.00	0.903	0.00
DUMP	0.847	0.00	0.849	0.00
TAILGATE	0.076	0.15	0.076	0.15
NON	0.139	0.02	0.140	0.02
T	0.381	0.01	0.381	0.01
NON×T	0.064	0.64	0.065	0.64
Constant	5.435		5.878	
θ	0.120	0.00	0.121	0.00
λ			1.036	0.00
log likelihood	−6,955		−6,955	
F-value (p-value)	1,845	(0.00)	1,845	(0.00)
likelihood ratio tests (1)	1,371.71	(0.00)	1,349.24	(0.00)
likelihood ratio tests (2)	16.47	(0.00)	25.23	(0.00)
likelihood ratio tests (3)	813.52	(0.00)	813.53	(0.00)

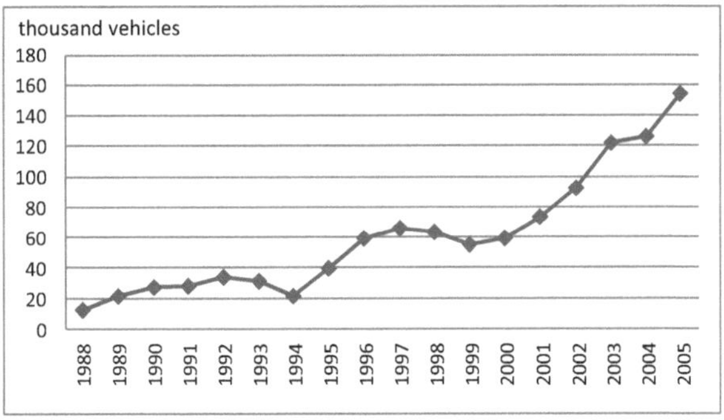

Fig. 4.3 Commercial exports of trucks

The coefficient on the interaction term, *COMPLYING* (whether a vehicle complies with VTR) and *NON* (whether a vehicle was sold outside the regulated areas) is found to be insignificant. This means that outside the regulated areas, vehicles subject to VTR are substitutes for those complying with VTR.

The coefficients on *FWDRIVE*, *AT*, *D-AIRBAG*, *PWINDOWS*, and *COMPLYING* are found to be positive and significant at the conventional level. The results indicate that vehicles are also priced higher when the following specifications are equipped: *RV*, *REEFER*, *BUCKET*, *CRANE* and *DUMP*. It was found, however, that the coefficient on *PINSPC* (whether a vehicle is periodically inspected) is negative and significant at the conventional level.

Overall, our results showed that prices of secondary vehicles outside the regulated areas were not changed significantly by the introduction of VTR. That is, the outflow of retired vehicles from the regulated to unregulated areas was not considerably large. This, in turn, suggests the possibility that many of the banned vehicles have been exported overseas.

Based on Fuse and Kashima (2007b), Fig. 4.3 was created to illustrate the volume of commercial exports of trucks. As shown in the figure, the number of exports rose sharply from 2000 and continually increased in 2003 when the first VTR ban became effective. In that year, the number of exports is almost the same as that of new standard trucks sold in the country. This may imply that retired trucks were exported overseas. VTR, which was meant to control domestic emissions, may negatively affect emissions reductions in developing countries overseas.

4.6 Conclusion

This chapter has estimated the economic impact of VTR on secondary vehicle markets in unregulated areas. Our results showed that secondary vehicle prices outside the regulated areas have not been changed significantly by the introduction of VTR. That is, the outflow of banned vehicles was not large enough to influence secondary vehicle prices in unregulated areas.

Meanwhile, the export volume of secondary vehicles is found to have increased greatly after the implementation, suggesting the possibility that banned vehicles have rather been exported overseas. This exemplifies a case of the pollution haven hypothesis; exporting polluting vehicles to developing countries for the sake of complying with regulations may be hindering emissions reductions in the importing countries.

In order to comprehensively evaluate the effectiveness of an environmental regulation, one needs to consider its potential effect on areas that are not directly regulated. This is especially so if a regulation controls the use of particular goods, like vehicles, whose secondary markets are well established. As we have seen in the case of VTR, even if a regulation intends to control domestic emissions, it can influence emissions in other countries. Hence, when drafting an environmental regulation on goods, it is essential to closely observe its secondary markets and predict in what way and to what extent the regulation may influence those markets.

References

Alberini A, Harrington W, McConnell V (1996) Estimating an emissions supply function from accelerated vehicle retirement programs. Rev Econ Stat 78(2):251–265

Dill J (2004) Estimating emissions reductions from accelerated vehicle retirement programs. Transport Res Part D 9(2):87–106

Fuse M, Kashima S (2007a) The effect of exhaust emissions regulation for vehicles on the export of used trucks and auto parts (in Japanese). J Jpn Soc Waste Manage Experts 18(1):20–29

Fuse M, Kashima S (2007b) Estimation of export volume for end-of-life vehicles from Japan (in Japanese). J Jpn Soc Waste Manage Experts 18(5):305–313

Iwata K, Arimura TH (2009) Economic analysis of Japanese air pollution regulation: an optimal retirement problem under vehicle type regulation in the NOx–particulate matter law. Transport Res Part D 14(3):157–167

Japan Automobile Manufacturers Association (JAMA) (2005a) Standard truck market survey 2004 (in Japanese). JAMA, Tokyo

JAMA (2005b) Small truck market survey 2004 (in Japanese). JAMA, Tokyo

Lin J, Chen C, Niemeier DA (2008) An analysis on long term emission benefits of a government vehicle fleet replacement plan in northern Illinois. Transportation 35(2):219–235

Lumbreras J, Valdés M, Borge R, Rodríguez ME (2008) Assessment of vehicle emissions projections in Madrid (Spain) from 2004 to 2012 considering several control strategies. Transport Res Part A 42(4):646–658

Matas A, Raymond JL (2009) Hedonic prices for cars: an application to the Spanish car market, 1981-2005. Appl Econ 41(22):2887–2904

Meese RA, Wallace NE (1997) The construction of residential housing price indices: a comparison of repeat-sales, hedonic-regression, and hybrid approaches. J Real Estate Financ Econ 14 (1–2):51–73

Ministry of Economy, Trade and Industry (2006) Approaches to encourage appropriate home appliance disposal in local governments (in Japanese). http://www.env.go.jp/council/former2013/03haiki/y0311-05/mat02_3.pdf. Accessed 15 May 2015

Ministry of Internal Affairs and Communications (2007) Status of trial implementation of regulatory impact analysis (RIA) (in Japanese). http://www.soumu.go.jp/main_content/000076915.pdf. Accessed 15 May 2015

Ministry of the Environment (2005) Vehicle type regulation under the automobile NOx-PM act (in Japanese). http://www.env.go.jp/air/car/pamph/. Accessed 15 May 2015

Ministry of the Environment (2007) Amendment of the automobile NOx-PM act (in Japanese). http://www.env.go.jp/air/car/pamph_kaiseihou/full.pdf. Accessed 15 May 2015

National Cooperative Highway Research Program (1997) Improving transportation data for mobile source emission estimates. Transportation Research Board, Washington, DC

Reis HJ, Silva JMCS (2006) Hedonic prices indexes for new passenger cars in Portugal (1997-2001). Econ Model 23(6):890–908

Rosen S (1974) Hedonic prices and implicit markets: product differentiation in pure competition. J Polit Econ 82(1):34–55

Sano M (2008) Genealogy of automotive technology innovation and technology policy in Japan (in Japanese). In: Miyoshi H, Tanishita M (eds) Technological innovation in the automotive industry and economic welfare. Hakuto Shobo, Tokyo, pp 1–36

Van Dalen J, Bode B (2004) Quality-corrected price indices: the case of the Dutch new passenger car market, 1990-1999. Appl Econ 36(11):1169–1197

Chapter 5
An Economic Welfare Analysis of the 1,000-Yen Expressway Discount

Abstract This chapter presents an economic welfare analysis of the 1,000-Yen Expressway Discount program (100 yen ≒ 1 US dollar) that was put into effect in March 2009 using data from the Tomei Expressway collected during the long holidays in spring of the same year. The new discount system was criticized because it might result in increased CO_2 emissions and thus adversely impact the environment. The analysis also examines the effects of the discount program on the Tokaido Shinkansen as an alternative means of transportation. We calculate the costs and benefits of the newly introduced discount system by comparing economic data for the same period in 2008 and 2009. According to our estimation, the costs incurred as a result of the new discount system exceeded the benefits by approximately 71–477 million yen. This result suggests that the discount program is ineffective as an economic policy in the case of the Tomei Expressway. The largest contributor to the net loss was the sharp increase in traffic congestion induced by the new program. Based on our findings, we recommend that different discount rates be set for different routes or time periods depending on the degree of congestion instead of the uniform discount rates currently applied throughout Japan.

Keywords Expressway Toll Discount • Greenhouse Gas Emissions • General Equilibrium Approach • Congestion • Ex-post Evaluation • Tomei Expressway

5.1 Introduction

The special holiday-discounted expressway toll rates, called the "1,000-Yen Expressway Discount," went into effect on March 28, 2009.[1] Under this new discount system, a 50 % discount was applied to almost all expressway toll rates for light and standard-sized cars with an electric toll collection (ETC) card during weekends and holidays. In addition, the maximum toll was capped at 1,000 yen.

[1] 100 yen approximately equals to 1 US dollar.

T.H. Arimura, K. Iwata, *An Evaluation of Japanese Environmental Regulations*,
DOI 10.1007/978-94-017-9947-8_5

Thus, long-distance trips using expressways, such as drives from Tokyo to Nagoya or Osaka, cost 1,000 yen,[2] representing a considerable discount. For example, the toll for driving 400 km was 8,300 yen before the discount and only 1,000 yen with the discount, representing an 88 % reduction in toll fees.

This new discount made headlines because the 1,000-Yen Expressway Discount had substantial effects on everyday life. The new discount system attracted considerably more attention when it was extended to weekdays during the long spring holiday season from April 25 to May 6 in 2009. The new toll rates were also applied to the weekdays of August 6–7 and 13–14 during the Obon period[3] in the same year.

Before the introduction of the 1,000-Yen Expressway Discount program, there had been a variety of toll discounts, such as the midnight discount, in which a 50 % discount was applied from midnight to 4 a.m. However, the new toll rate program with the maximum charge capped at 1,000 yen offered a much higher discount than the other special discounts.

This expressway discount program was one of the economic measures adopted by the government. The policy was intended to promote the use of expressways by lowering the costs of expressway use, to help increase the distribution of goods and services, and to stimulate local economies along the highway and thereby ultimately stimulate the Japanese economy.

However, heavier expressway traffic as a result of discounts does not always favor economic welfare. An increase in the use of expressways increases fuel consumption and leads to increased production of air pollutants, such as NOx and PM and increased CO_2 emissions, which are the source of greenhouse gas effects. Furthermore, increased use of expressways may result in traffic congestion in many places where traffic jams have never occurred. An expressway toll discount can also be viewed as a variant form of the policy of selectively providing subsidies to promote the use of privately owned cars among various transportation systems. Thus, alternative transportation providers, such as railway companies, may face a decrease in demand due to the substitution effect.

As mentioned above, the 1,000-Yen Expressway Discount could have both negative and positive effects. In fact, a report published by the Institute of Transportation Economics on October 2, 2009 (Institute of Transportation Economics 2009) estimated that the new expressway discount would increase the amount of annualized CO_2 emissions by 2,040 thousand tons. Therefore, to determine whether the newly introduced expressway discount program had net positive effects on the economy rather than net negative effects, we need to consider not only the benefits of the lower tolls (an increase in consumer surplus) but also economic losses (an increase in external costs), such as increased environmental costs.

---

[2] The new discounts were not applicable to the section of the Tomei Expressway between Tokyo IC and Atugi IC because this section passes through the metropolitan area and its suburbs. We could not obtain traffic data for this section and had to assume that the discounts had been applied to it. See the following NEXCO East webpage for the metropolitan area and its suburbs: http://www.driveplaza.com/etc/holiday_daytime_discount/about.html#gaiyo.

[3] This is the Japanese summer holiday week during which many people visit their hometowns.

The Ministry of Land, Infrastructure, Transportation, and Tourism (MLIT: Japanese Department of Transportation) estimated that the new discount program would increase spending on travel and tourism by 730 billion yen and reduce transportation costs by 200 billion yen over 2 years. According to the calculations of the MLIT, the economic ripple effects would amount to 1,700 billion yen. At a press meeting on August 10, 2009, however, the administrative vice minister of the MLIT stated that these estimations did not consider the negative impact caused by the toll discount, such as the adverse effect on the environment. In summary, it remains unclear whether the 1,000-Yen Expressway Discount was truly beneficial for the economy.

Even if the MLIT's estimation on the positive side was correct, whether the new expressway discount program was effective as an economic policy should be questioned if its negative effects on the environment, road traffic, and other transportation systems exceeded the positive effects. Specifically, if the 1,000-Yen Expressway Discount resulted in environment-related external costs that exceeded the economic benefits, the program has no economic rationale as a viable policy.

Using the economic data related to the Tomei Expressway during the so-called "Golden Week" in 2009, a long spring holiday season of April 25 to May 6, we assess the net economic impact of the special holiday expressway toll discount program and discuss whether the newly introduced pricing system played a positive economic role in promoting social welfare.

This chapter differs from Chaps. 2 and 3, in which we evaluate the Automobile NOx-PM Act in the following respects. First, in this chapter, we examine ex post data. In Chaps. 2 and 3, we examined the regulation using information available before the regulation was implemented. Therefore, in one sense, we analyzed a hypothetical scenario. In contrast, in this chapter, we use ex post data, observed after the expressway discount was introduced.

Second, we use a general equilibrium approach. More specifically, this chapter is different from Chaps. 2 and 3 in the sense that we examine the impact of the new price rule in other markets as well. This chapter focuses not only on the Tomei Expressway but also on the Tokaido Shinkansen as an alternative transportation system.

This chapter is organized as follows. Sections 5.2 and 5.3 explain the concepts of an economic welfare analysis, including average costs, demand under general equilibrium, and average social costs. Various costs that resulted from the new pricing system are estimated in Sect. 5.4, and the costs and benefits with respect to the Tomei Expressway as a whole are calculated in Sect. 5.5. Section 5.6 presents the cost-benefit analysis for the Tokaido Shinkansen and aggregates the total costs and benefits that the Tomei Expressway and the Tokaido Shinkansen generate. Finally, Sect. 5.7 concludes this chapter.

5.2 Economic Welfare Analysis Using Average Costs

Chapter 1 presents an economic welfare analysis using marginal costs; however, it is not always easy to estimate marginal costs. Instead, we use average costs to measure economic welfare in this chapter. We first explain how to measure economic welfare using average costs.

Figure 5.1 presents the demand, supply, and average cost curves for a particular good. As discussed in Chap. 1, the supply curve is identical to the marginal cost curve. First, we conduct a welfare analysis using marginal costs. The area of $\square OQ^AAF$ represents the total cost at equilibrium where the amount Q^A is produced. Hence, the area of $\triangle FAP^A = \square OQ^AAP^A - \square OQ^AAF$ represents the producer surplus.

The demand curve represents the consumer marginal utility, and the area of $\square OQ^AAE$ is equal to the maximum amount of money that consumers are willing to pay for that good. On the other hand, the area of $\square OQ^AAP^A$ represents the actual amount paid by consumers. Hence, the consumer surplus is calculated as $\triangle P^AAE = \square OQ^AAE - \square OQ^AAP^A$.

Thus, the social economic surplus, defined as the sum of the consumer and producer surplus, is expressed as $\triangle FAE = \triangle FAP^A + \triangle P^AAE$. A similar discussion is presented in Chap. 1 of this book and can be found in introductory microeconomic textbooks.[4]

Next, we turn our attention to understanding how to measure the producer surplus using average costs. The average cost is defined as the total cost divided by the number of units produced. In our example, the equilibrium total production cost is calculated as the area of $\square OQ^ABD$ in Fig. 5.1. The producer revenue is $\square OQ^AAP^A$, which is identical to the producer revenue in the marginal cost

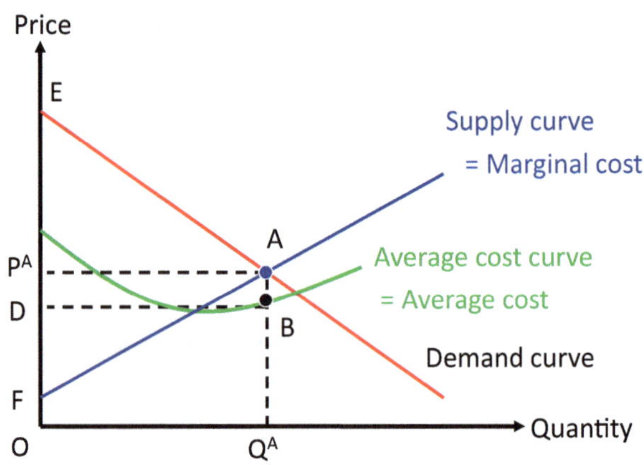

Fig. 5.1 Marginal cost and average cost

[4] Please see introductory microeconomic textbooks, such as Varian (1992).

approach. The producer surplus is represented by the area of □DBAP^A. Of course, the producer surplus remains unchanged whether the marginal or average cost is used to measure economic welfare. That is, △FAP^A = □DBAP^A. In contrast, △P^{A}AE represents the consumer surplus in both approaches. Therefore, the social economic surplus is equal to the area of □DBAE = □DBAP^A + △P^{A}AE in the average cost approach. Hence, the economic social welfare changes from △FAE to □DBAE in the average cost approach.

One of the advantages of the average cost approach is the convenience of its use in the evaluation of economic welfare (Boardman et al. 2010). The marginal cost curve must be estimated when measuring the surplus in the marginal cost approach. However, the estimation is typically not an easy task for many reasons. For instance, we can hardly imagine a firm that would be willing to disclose internal information related to its cost structure to the public. Thus, although econometric methods are well established, in reality, economists find it difficult to obtain the corporate cost structure data needed to conduct quantitative analysis.

In contrast to the marginal cost approach, the average cost approach does not require specification of the cost function of the firm. The only information required is relevant production and price data before and after the implementation of the policy to be assessed. In our study, this requirement means that we only need data on expressway traffic and tolls before and after the implementation of the 1,000-Yen Expressway Discount. Our focus is on the social costs and benefits related to expressways. Travelling on expressways involves fuel consumption and other expenses and incurs external costs, such as environmental damage from greenhouse gas and air pollutant emissions. Given these negative effects, we must collect data on pollution and use the concept of the average social cost. The next section presents this concept in detail.

More precise analysis requires the estimation of the average cost function. Still, in comparison with the marginal cost approach, the average cost approach is less time consuming and convenient for evaluating policy effects. Thus, a social welfare analysis using the average cost approach has the advantage of being convenient for use in conducting policy assessments.

5.3 Generalized Equilibrium Demand for the Tomei Expressway and Social Cost

5.3.1 Changes Observed in the Tomei Expressway

This section presents an analysis conducted using a generalized equilibrium demand curve. This approach employs ex post information about what occurred after the 1,000-Yen Expressway Discount policy was established. We compare data from 2008 with data from 2009, when the discount was introduced. By examining the changes in costs and benefits from 2008 to 2009, we are able to estimate changes in social economic welfare.

In our analysis, we assume that the discount policy is responsible for all changes observed after its implementation. Economic conditions and other factors may have changed during the period and may have had some effects on the costs and benefits observed. We discuss this issue later in this chapter. In addition, we consider the impact of the new expressway pricing system on other transportation systems, such as the Tokaido Shinkansen bullet train. The impact on the Tokaido Shinkansen can be substantial because the Tomei Expressway and Tokaido Shinkansen connect Tokyo to Nagoya in parallel. Section 5.6 addresses this issue.

Let us look at what happened to the Tomei Expressway during the "Golden Week" in 2009 when the 1,000-Yen Expressway Discount was implemented. First, a sharp decrease in toll revenues was observed as a result of the new discount program. Second, the lower tolls increased the number of traveling cars. This increased traffic simply reflects the fact commonly observed across markets that lower prices result in increased demand. Third, the increased traffic volume caused more frequent traffic congestion and longer queues. Fourth, travel times to destinations increased due to of the increased number of traffic jams and longer lines of cars, with consequent increases in petrol consumption, emissions of air pollutants and greenhouse gases, and the number of road accidents. Figure 5.2 summarizes various effects of the new expressway toll discount program. The colored items represent the costs incurred as a result of the new discount toll system with which we are concerned. Welfare losses for each of the colored items are calculated in this chapter.

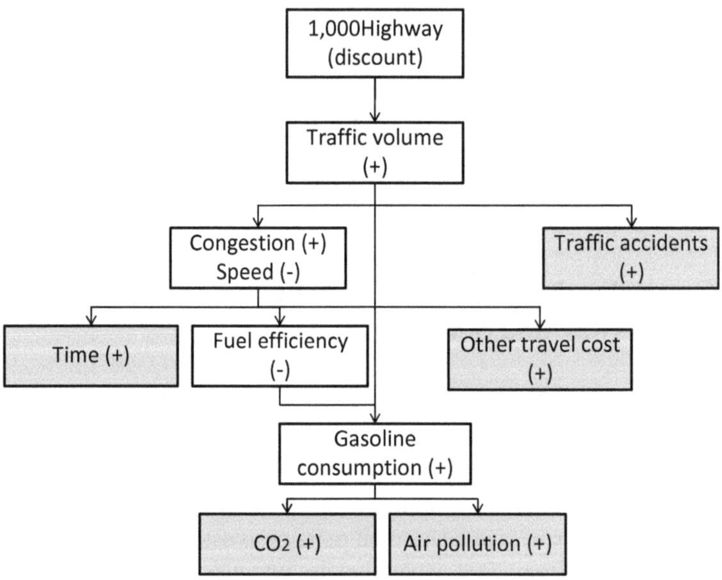

Fig. 5.2 Effects of the discount program on the market of the Tomei Expressway

5.3.2 Generalized Cost and Benefits for the Tomei Expressway due to the 1,000-Yen Expressway Discount

We consider a variety of services generated by the use of the Tomei Expressway as a single service. For simplicity, we suppose that consumers consume the Tomei Expressway service by driving their cars. Consumers pay a toll for receiving the service. In addition, consumers bear a wide range of costs for the service. These costs are set out below.

First, traveling from a starting point to a destination (in our analysis, the travel time between two toll gates) incurs the opportunity cost of the alternative use of time. Economics considers time as money. A consumer has choices and can earn a certain amount of money if she or he chooses to work instead of driving on an expressway. This potential income is counted as an opportunity cost.

Second, cars do not run without fuel, so consumers must purchase fuel to drive cars. They also need to pay for wear-and-tire items, such as oil and tires, as well as maintenance and repairs. An automobile is a durable good; thus, the value of a car decreases as the consumer travels more. If consumers drive their cars on the Tomei Expressway, they must consider depreciation cost.

The total of the costs involved in the use of the expressway service is called the generalized cost (Boardman et al. 2010). The generalized cost represents the amount of money that consumers are willing to pay for a particular good or service, and it can be considered as a type of demand curve. Given the above discussion, the generalized cost is different from the standard notion of cost and includes all relevant costs associated with travelling. The generalized cost is given in Eq. (5.1). Other traveling cost covers expenses for oil, tires, maintenance, and depreciation.

$$\begin{aligned}\text{Generalized cost} = {} & \text{toll} + \text{time cost} + \text{fuel cost inclusive of tax} \\ & + \text{other traveling costs}\end{aligned} \tag{5.1}$$

In Fig. 5.3, we use the concept of the generalized cost and review how much consumer surplus and social benefit changed as a result of the 1,000-Yen Expressway Discount. P^A represents the generalized cost for expressway service in 2008, the year before the discount, and Q^A represents the number of cars that used the expressway in that year.

Tolls fell sharply in 2009 due to the implementation of the new discount program. As we mentioned, the increasing frequency and severity of traffic congestion led to longer travel times. Thus, the benefits of the toll discounts were partly offset by longer travel time with respect to generalized cost. P^B represents the generalized cost for expressway service in 2009, and Q^B represents the number of cars that used the expressway in that year.

This result means that the generalized cost moved from A to B along the general equilibrium demand curve with the new pricing system. Data on the generalized cost and traffic volumes corresponding to the two points A and B on the demand

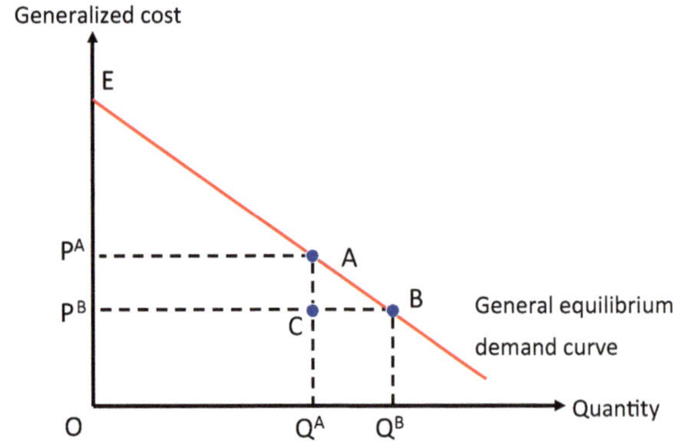

Fig. 5.3 General equilibrium demand curve

curve can be easily obtained; however, we must make several assumptions to extrapolate the demand curve with these two points. For simplicity, we assume that the demand curve takes the form of a straight line that connects A and B.

This straight line is called a general equilibrium demand curve. The new expressway discounts increased the traffic volume from Q^A to Q^B, which reflects the fact that some users of other transportation systems, such as Shinkansen, used the expressways instead because of the decrease in tolls. Thus, the downward-sloping demand curve in Fig. 5.3 considers the effects of lowering tolls on other markets.

In the year 2008, before the discounts were put in place, the generalized cost and traffic volume were P^A and Q^A. Thus, $\triangle P^AAE$ represents the consumer surplus for the year. In 2009, the consumer surplus changed to $\triangle P^BBE$ due to the decrease in the generalized cost to P^B and the increase in the traffic volume to Q^B. Thus, the net increase in the consumer surplus driven by the 1,000-Yen Expressway Discount is represented by the area of $\square P^BBAP^A$ in Fig. 5.3.

Next, we define the benefits from the expressway service. The benefits from the 1,000-Yen Expressway Discount can be defined as an increase in consumers' willingness to pay. This increased willingness to pay corresponds to an increase in the area below the generalized demand curve. The relation between the consumer surplus and benefit is shown below:

$$\text{Benefit} = \text{increase in the consumer surplus} - \text{increase in spending} \quad (5.2)$$

We illustrate the benefit of the discount in Fig. 5.3. As we explain, $\square P^BBAP^A$ represents the increase in consumer surplus, whereas $\square OQ^AAP^A$-$\square OQ^BBP^B = \square P^BCAP^A$-$\square Q^AQ^BBC$ represents the increase in spending. Thus, Eq. (5.2) indicates that the benefit becomes $\square Q^AQ^BBA$, which is equal to the increase in consumers' willingness to pay. This increased willingness to pay implies an increase in utility for consumers.

5.3.3 Average Social Cost and the Cost of the 1,000-Yen Expressway Discount for the Tomei Expressway

The generalized cost represents a monetary amount of cost that expressway service users bear. This section discusses the social costs related to the expressway service market. As introductory microeconomic textbooks explain, the social cost is the sum of the private cost and external cost. The average social cost is defined as shown in (5.3) below.

$$\text{Average social cost} = \text{private average cost} + \text{average external cost} \tag{5.3}$$

Following Boardman et al. (2010), we assume that the social cost consists of the time cost, the fuel cost after tax, and other driving costs and external costs, as shown in (5.4). We exclude fuel tax[5] from the social cost because the fuel tax collected is ultimately spent in various forms for the public and is considered an income transfer in which only the owners of the money change.

In this chapter, we consider climate change, air pollution, and traffic accidents to be external costs. Typically, the cost of time is regarded as an external cost. However because the 1,000-Yen Expressway Discount caused a substantial increase in the number of occurrences of traffic congestion and in travel times, we treat the cost of time separately so that we can investigate it in detail.

$$\begin{aligned}&\text{Average social cost} = \text{time cost} + \text{fuel cost} + \text{other cost} + \text{average external cost}\\&\text{Average external cost} = \text{climate change} + \text{air pollution} + \text{traffic accident}\end{aligned} \tag{5.4}$$

Figure 5.4 adds the average social cost curve to Fig. 5.3 with the generalized demand curve. We see the extent to which economic welfare was changed by the introduction of the 1,000-Yen Expressway Discount using Fig. 5.4. As in Fig. 5.3, Q^A and P^A represent the traffic volume and generalized cost before the new discount system. Suppose that the traffic volume increased to Q^B and the generalized cost decreased to P^B after the new discount system was introduced. The change in the benefit associated with the new discount is represented by □Q^AQ^BBA. What about the average social cost? Before the introduction of the new discount system, the level of the average social cost was at P^C and the social cost represented by □O Q^AC P^C = line segment O Q^A × line segment Q^AC.

However, after the new discount system was introduced, the average driving speed became slower as travel times increased, and fuel consumption increased due to frequent traffic congestion. As a result, the average social cost increased substantially to P^D, and the increase in the traffic volume led to a large increase in the total social cost, represented by □O Q^BD P^D.

[5] The fuel tax is imposed on gasoline use in Japan. Please see Chap. 7 for the exact amount and the structure of the tax.

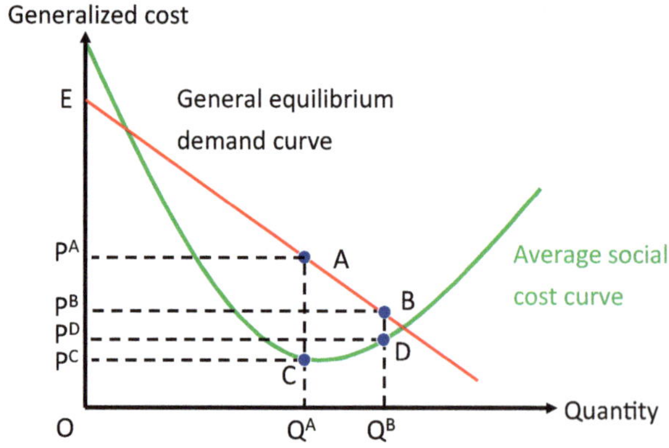

Fig. 5.4 Social cost generated by the new discount system

Table 5.1 Lists of costs

	Contents	Detail
Generalized cost	Expressway toll	
	Time cost	
	Fuel cost and tax	
	Other driving cost	Oil, grease, tires and tubes expense; maintenance and repairing cost; depreciation cost
Social cost	Time cost	
	Fuel cost	
	Other driving cost	Oil, grease, tires and tubes expense; maintenance and repairing cost; depreciation cost
	Average external cost	Climate change; air pollution; traffic accident

Thus, an increase in the social cost generated by the 1,000-Yen Expressway Discount takes the shape of $P^{C}CQ^{A}Q^{B}DP^{D} = \square OQ^{B}DP^{D}$ (social cost after discount) – $\square OQ^{A}CP^{C}$ (social cost before discount).

The purpose of the analysis presented in this chapter is to evaluate the toll discount policy to compare the increase in costs (the shape $P^{C}CQ^{A}Q^{B}D\ P^{D}$) with the increase in benefits ($\square Q^{A}Q^{B}BA$) and determine which effect is dominant. Please refer to Boardman et al. (2010) for a detailed explanation of the analysis using average social cost and generalized cost.

We are now prepared to calculate the costs listed above in the generalized cost Eq. (5.1) and the social cost Eq. (5.4). Before calculating these costs, we examine the differences in the data for the Golden Weeks of 2008 and 2009 for the Tomei Expressway. Table 5.1 presents these costs.

5.4 Changes in Costs

5.4.1 *Expressway Toll System*

First, let us look at the expressway toll system. Drivers of standard-sized cars are charged 24.6 yen/km for traveling on expressways. In addition, another 150 yen is charged for covering fixed costs regardless of the traveling distance, and discounts are applied to long-distance drives.

A 25 % discount is applied to distances over 100 km. For example, suppose that a driver drives 150 km on expressways. The driver receives a 25 % discount for 50 km of that trip. A 25 % discount is applied to travel distances between 100 and 200 km, and a 30 % discount is applied to distances over 200 km. The new toll rate program, with a 50 % discount and tolls capped at 1,000 yen, was introduced in 2009. Table 5.2 compares the tolls in years 2008 and 2009 by distance. The details of the toll system are presented in Appendix 5.1.

A decrease in expressway tolls means a decrease in transportation cost, which increases the benefit to users. To assess economic welfare, we must calculate the benefits by travel distance. However, we face a data constraint: detailed data on travel distances are not available. Therefore, we divide travel distances using the Tomei Expressway into seven categories, presented in Table 5.2, for the purposes of our analysis. For reference, we assign a number to each category. For example, the driving distance categories of 25 and 75 km are denoted as categories 1 and 2, respectively. The numbered categories appear in column one in Table 5.2.

5.4.2 *Traffic Volume*

As extensively reported by the media, a substantial increase in traffic volume was observed due to the 1,000-Yen Expressway Discount. According to NEXCO

Table 5.2 Expressway tolls

Category	Traveling distance (km)	2008 Expressway toll per car (yen)	2009 Holiday discounted expressway toll per car (yen)
1	25	800	400
2	75	2,100	1,000
3	125	3,250	1,000
4	175	4,200	1,000
5	250	5,600	1,000
6	350	7,400	1,000
7	400	8,300	1,000

Central (2009), the traffic volume on the Tomei Expressway was 88,000 vehicles per day during the "Golden Week" period of April 25 to May 6 in 2008 and rose to 93,100 vehicles per day, or nearly 6 % higher, during the same period in 2009 with the new expressway discount. The numbers of vehicles that traveled on the Tomei Expressway during the 12-day-long Golden Week of 2008 and 2009 were 1,056,000 and 1,117,200, respectively.

NEXCO Central (2009) also reported the number of cars that passed through the Nagoya Interchange toll gates by driving distance per day (Table 5.3). A total of 32,400 cars per day passed through the Nagoya Interchange toll gates in 2008. The number of cars travelling 50 km or less and the number traveling 51–100 km were 18,500 (57 %) and 6,000 (19 %), respectively. Comparing the numbers of cars by the driving distance in 2008 and 2009, the number of cars travelling 51–100 km decreased slightly from 2008 to 2009, whereas the numbers in all other categories increased.

Table 5.3 is based on the NEXCO Central report, but Table 5.2 uses different categories for travel distances. To use the travel distance classification shown in Table 5.2, we must make some adjustments to the classification in Table 5.3 so that we can compare it to Table 5.2. For this purpose, we assume that the number of cars is uniformly distributed over the range of travelling distances in each category. For example, we put the 0- to 50-km category in Table 5.3 into the 25-km category in Table 5.2 and the 51- to 100-km category into the 75-km category. For cars travelling farther than 400 km in Table 5.3, there is no upper limit, and we take this travel distance as the 400-km category in Table 5.2. Please see Appendix 5.2 for details. Table 5.4 presents our estimation of the number of cars by travel distance.

Due to the availability of data and for the sake of simplicity in our approach, we use the data broken down by travel distance to measure economic welfare. To do so, we divide travel distances, starting from 25 km, by 50 km. For each category, we calculate costs and benefits and aggregate them. Policy assessment is typically subject to time constraints, and timeliness is a key factor. Therefore, it is important to use a convenient approach, such as the one we use in this chapter.

Table 5.3 Number of cars at the Nagoya Interchange toll gates (Golden Week)

Traveling distance (km)	2008	2009	2008	2009
	Cars per day	Cars per day	Ratio	Ratio
~50	18,500	18,900	0.57	0.45
~100	6,000	5,900	0.19	0.14
~150	1,300	2,600	0.04	0.06
~200	1,500	3,300	0.05	0.08
~300	2,200	5,200	0.07	0.12
~400	2,700	5,400	0.08	0.13
over 400	200	600	0.01	0.01
Total	32,400	41,900	1.00	1.00

Table 5.4 Number of cars (Tomei Expressway, Golden Week)

Category	Traveling distance (km)	2008 Cars per day	2009 Cars per day	2008 Traffic volume	2009 Traffic volume
1	25	50,247	41,995	602,963	503,940
2	75	16,296	13,110	195,556	157,315
3	125	3,531	5,777	42,370	69,325
4	175	4,074	7,332	48,889	87,989
5	250	5,975	11,554	71,704	138,650
6	350	7,333	11,999	88,000	143,983
7	400	543	1,333	6,519	15,998
	Total	88,000	93,100	1,056,000	1,117,200

5.4.3 *Traffic Congestion*

As the media reported, the new expressway toll rate policy resulted in a sharp increase in traffic and severe road congestion. We made an inquiry to NEXCO Central and obtained information and data concerning traffic congestion on the Tomei Expressway during the Golden Week period in 2008 and 2009. This information and data are termed "CI (congestion information)" hereinafter. However, the data and information available to us are limited to periods of traffic congestion with queues longer than 10 km. CI includes the place, date, time, peak time, and longest queue length for each congestion occurrence.[6]

According to CI, the Tomei Expressway was congested 39 times during the Golden Week in 2008, and this number more than doubled to 82 during the same period in 2009. Similarly, the total queue length was 837.9 km in 2008 and more than doubled to 1,709.0 km in 2009. However, the queue length per occurrence decreased slightly from 21.48 km in 2008 to 20.84 km in 2009.

CI recorded the starting time and peak time for each occurrence of congestion, but CI has no record of when the traffic congestion was dissolved. Therefore, we assume that the time intervals between the starting time and peak time and between the peak time and the end time are the same, and we calculate the duration of traffic congestion as (peak time – starting time) × 2 = duration. Based on this definition, we estimated that the total duration of congestion was 14,872 min in 2008 and 32,556 min in 2009. The total duration was 119 % greater in 2009 than in 2008.

From an economic perspective, traffic congestion causes time loss to individuals. Economics present the value of time in monetary units. When the MLIT builds a new road, the Ministry calculates the value of time savings in yen. According to the MLIT (2008a), the average time cost for a standard-sized car in our analysis is 40.1

[6] The press release by NEXCO Central presents information on traffic congestion for expressways as a whole. However, detailed information on congestion is limited to the five places where the five longest queues were observed.

yen/min. We multiply the increase in driving time caused by congestion by 40.1 yen/min to obtain the time cost per car as a function of travel distance.

5.4.4 *Time Cost Estimation*

The new toll discount system resulted in an increase in travel time and more frequent occurrences of traffic congestion. These consequences represent economic losses. In this chapter, we use the concept of expected queue length in traffic congestion multiplied by travel distance to estimate time cost (please see Appendix 5.3 for a detailed explanation of the expected queue length in traffic congestion). Suppose that a driver drove 400 km on the Tomei Expressway in 2008, with 82.28 km driven in congestion at a reduced speed and 317.72 km was driven at a normal speed. We assume that the driving speed is 40 km in congestion, based on the definition of traffic congestion, and that the normal driving speed is the legal speed of 80 km. The average travel time by driving distance is given as follows:

$$\begin{aligned}\text{Travel time} = &\{(\text{expected queue length})/(\text{speed in congestion})\}\\ &+(\text{driving distance} - \text{expected queue length})/(\text{normal speed})\end{aligned} \tag{5.5}$$

Table 5.5 presents the estimation of the travel time. The table shows that in 2008, it took approximately 113 min to drive 125 km, whereas it took 10 min longer, or approximately 123 min, in 2009. Congestion led to a substantially longer travel time. Tables 5.5 and 5.6 presents the results of translating increases in travel time into time costs based on the basic unit cost determined by the MLIT (2008a) (see Appendix 5.4 for details).

Table 5.5 Travel time (Tomei Express during the Golden Week)

Category	Traveling distance (km)	2008 Traveling time per car (minute)	2009 Traveling time per car (minute)
1	25	22.6	24.6
2	75	67.8	73.9
3	125	113.0	123.1
4	175	158.3	172.4
5	250	226.1	246.3
6	350	316.5	344.8
7	400	361.7	394.1

Table 5.6 Travel Cost (Tomei Express during the Golden Week)

Category	Traveling distance (km)	2008 Time cost per car (yen)	2009 Time cost per car (yen)
1	25	906.5	987.6
2	75	2,719.6	2,962.8
3	125	4,532.7	4,938.0
4	175	6,345.7	6,913.2
5	250	9,065.4	9,876.1
6	350	12,691.5	13,826.5
7	400	14,504.6	15,801.7

5.4.5 *Fuel Efficiency*

Frequent traffic congestion causes inefficient fuel consumption. The MLIT (2008b) uses the following relation between driving speed and fuel consumption. We calculate the fuel consumption in congestion and in normal conditions based on this equation.

$$\text{Fuel consumption(cc/km)} = 829.3/(\text{speed}) - 0.9(\text{speed}) + 0.0077(\text{speed})^2 + 64.1 \tag{5.6}$$

If we substitute 40 km/h for the speed in congestion and 80 km/h for the normal speed into the speed term in (5.6), we obtain fuel efficiencies of 16.35 km/litter in congestion and 19.33 km/litter in normal conditions.

5.4.6 *Fuel Consumption*

A larger traffic volume and longer travel distance result in increased fuel consumption. In addition, traffic congestion causes inefficient fuel consumption. Thus, based on the distances driven at a normal speed and at a slow speed in traffic congestion, we can estimate the increase in fuel consumption.

The calculation of driving distance in congestion and under normal conditions is presented in Appendix 5.2. We have also calculated fuel efficiency in Sect. 5.4.5. Using these results and Eq. (5.7) below, we can estimate fuel consumption as a function of travel distance. Table 5.7 presents the results of our calculation for fuel consumption per car. We multiply the fuel consumption for each travel distance group by the number of cars in each group and aggregate the results to obtain the overall fuel consumption.

Table 5.7 Gasoline consumption (Tomei Express during the Golden Week)

Category	Traveling distance (km)	2008 Fuel consumption per car (litter)	2009 Fuel consumption per car (litter)
1	25	1.34	1.37
2	75	4.03	4.1
3	125	6.71	6.84
4	175	9.39	9.57
5	250	13.42	13.67
6	350	18.79	19.14
7	400	21.47	21.88

$$\begin{aligned}(\text{Fuel consumption}) = {} & (\text{driving distance in congestion})/(40\ \text{km/h fuel efficiency}) \\ & + (\text{driving distance under normal condition})/ \\ & (80\ \text{km/h fuel efficiency})\end{aligned} \tag{5.7}$$

5.4.7 *Other Mileage Costs*

Drivers have to bear a variety of costs in addition to fuel and time costs to drive their car. Typical examples are expenses for oil and grease, tires and tubes, maintenance and repair, and depreciation. We estimate these "other costs" in this section.

To calculate these costs, we refer to various formulae and basic units provided in the MLIT (2008b). For the cost related to oil and grease, we calculate the amount per kilometer of driving distance at normal and slow speeds in congestion based on Eq. (5.8). Substituting these numbers into (5.8), we obtain values of 0.082 yen/km under normal conditions and 0.097 yen/km in congestion.

$$\begin{aligned}(\text{Oil and grease expense})\,(\text{yen/km}) = {} & 1.285 - 0.00146(\text{speed}) \\ & + 0.0000124(\text{speed})^2 + 0.1032\end{aligned} \tag{5.8}$$

Similarly, we calculate the cost of tires and tubes using Eq. (5.9) and obtain values of 0.770 yen/km under normal conditions and 0.417 yen/km in congestion.

$$(\text{Cost for tires and tubes})\,(\text{yen/km}) = 0.9/[-0.02463 \times (\text{speed}) + 3.319] \tag{5.9}$$

Maintenance and repair costs and depreciation costs are not considered to be affected by the driving speed. Therefore, we assume a value of 2.17 yen/km for maintenance and repair and 2.37 yen/km for depreciation.

Table 5.8 Other traveling costs (Tomei Express during the Golden Week)

Category	Traveling distance (km)	2008 Other mileage cost per car (yen)	2009 Other mileage cost per car (yen)
1	25	133.1	132.2
2	75	399.2	396.5
3	125	665.3	660.8
4	175	931.5	925.1
5	250	1,330.6	1,321.6
6	350	1,862.9	1,850.2
7	400	2,129.0	2,114.5

Our estimated mileage by travel distance, excluding fuel cost, is presented in Table 5.8. Slower driving speeds result in lower costs for tires and tubes due to slower wear and tear. For this reason, other mileage costs were lower in 2009 than in 2008 due to the more frequent occurrence of traffic congestion in 2009.

5.4.8 External Cost: Environmental Cost

With respect to the 1,000-Yen Expressway Discount, the media and press reported rising concerns about its impact on the environment, such as a potential increase in CO_2 emissions. To assess the environmental effect of the new discount system and translate it into monetary units, we must estimate the possible adverse impact on the environment as an external cost.

We consider the issues of air pollution and climate change because automobiles are one of the main causes of these problems. The extent to which the 1,000-Yen Expressway Discount worsened air pollution and increased greenhouse gas emissions can be expressed in terms of increased fuel consumption if we use appropriate conversion rates. An increase in the number of traffic accidents has also been identified as a major external diseconomy concerning the use of cars. The media and press reported an increase in traffic accidents after the new pricing system was implemented.

Hence, in our welfare analysis, we assume that the external cost incurred as a result of increased automobile traffic consists of the costs of climate change, air pollution, and traffic accidents, as shown in Eq. (5.10). The cost of traffic congestion is excluded because it is included in the time cost, which we have already estimated.

$$\text{External cost} = \text{climate change} + \text{air pollution} + \text{traffic accident} \tag{5.10}$$

To measure external cost in yen, we use the following basic units cited in Kanemoto et al. (2006) and the MLIT (2008b):

Table 5.9 Negative externalities (air pollution, climate change, traffic accidents: Tomei Expressway, Golden Week)

Category	Traveling distance (km)	2008 External cost per car (yen)	2009 External cost per car (yen)
1	25	335.1	335.8
2	75	1,005.2	1,007.5
3	125	1,675.4	1,679.1
4	175	2,345.5	2,350.7
5	250	3,350.8	3,358.2
6	350	4,691.1	4,701.5
7	400	5,361.2	5,373.1

climate change cost: 19.3 yen/litter (Kanemoto et al. 2006)
air pollution cost: 9.9 yen/litter (Kanemoto et al. 2006)
traffic accident cost: 0.99 yen/km·vehicle·day (the MLIT 2008a)

Using these figures, we estimated the external cost incurred per car by travel distance during the 12 days of the Golden Week. Table 5.9 presents our results. The table illustrates that the external cost (or the environmental cost) increased in 2009.

5.5 The Cost and Benefits of the Tomei Expressway

5.5.1 Generalized Cost

One of the studies on economic welfare that considers the substitutability between expressways and public roads is Kanemoto et al. (2006). Their study presents an economic welfare analysis using the general equilibrium demand curve with average costs, including the spillover effects, such as the shift in demand from expressways to public roads and vice versa due to a decline in toll rates (see Sects. 5.2 and 5.3). Our study follows their approach.

In Sect. 5.4, we calculate several costs per car and by travel distance for the Tomei Expressway in 2008, the year before the new discount, and in 2009. The expressway tolls, time cost, fuel cost and tax, and other driving costs appear in Table 5.2, Table 5.6, Appendix 5.5, and Table 5.8, respectively. Based on these tables and Eq. (5.1), we calculate the generalized cost by travel distance. Table 5.10 presents our results.

Take category 3 in Table 5.10 as an example. The generalized cost per car for the Golden Week was 9,421.7 yen/car, and the traffic volume was 42,370 cars (Table 5.4). In 2009, when the new discount system was implemented, the cost decreased to 7,400.9 yen/car, while the traffic volume increased to 69,325 cars (Table 5.4).

Table 5.10 Generalized cost (Tomei Expressway, Golden Week)

Category	Traveling distance (km)	2008: median Generalized cost per car (yen)	2009: median Generalized cost per car (yen)
1	25	2,034.3	1,680.2
2	75	5,803.0	4,840.6
3	125	9,421.7	7,400.9
4	175	12,840.4	9,961.3
5	250	17,943.5	13,801.9
6	350	24,680.8	18,922.7
7	400	28,049.5	21,483.0

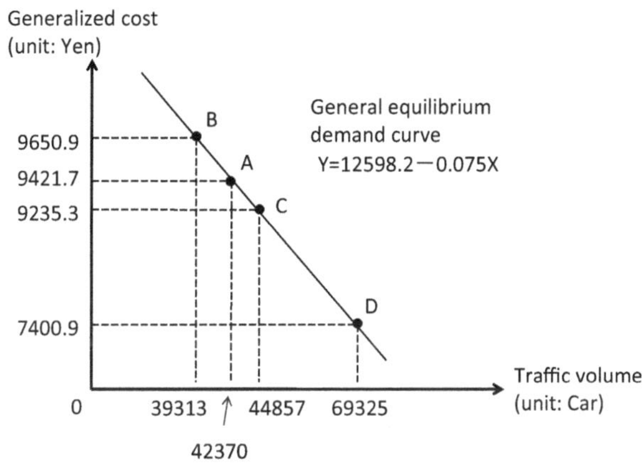

Fig. 5.5 Generalized demand function (category 3 in Tomei Expressway)

Here, we take the category 3 travel distance as an example and examine the relation between the generalized cost and the number of travelling cars in Fig. 5.5. The table reproduces Table 5.3 in Sect. 5.3, in which we explain the generalized cost approach, with the actual figures calculated in Sect. 5.4. Point A represents the generalized cost and traffic volume in 2008, and Point D represents the generalized cost and traffic volume in 2009. Assuming that the relationship between the generalized cost and traffic volume is linear, we can obtain the general equilibrium demand curve from the information given at points A and D. The general equilibrium demand function for category 3 is given below (5.11).

$$\text{Generalized cost} = 12,598.2 - 0.075 \times \text{traffic volume} \qquad (5.11)$$

We have to keep in mind that two major factors were primarily responsibly for the changes in the traffic volume. One was the newly introduced expressway toll discount, in which we are interested, and the second was a substantial decline in fuel prices, exclusive of tax, between 2008 and 2009. Therefore, we must exclude the changes brought about by the lower fuel prices exclusive of tax.

To address this issue, we assume that the level of fuel prices inclusive of tax remained unchanged from 2008 to 2009, and we consider two scenarios. To distinguish the generalized cost calculated from actual data for 2008 and the generalized cost in our scenarios, we call the former the median.

- Lower-Limit Scenario
  In this scenario, an increase in traffic congestion is assumed to be fully attributable to the new discounts. We assume that while the traffic volume in 2008 changed due to the same fuel prices, exclusive of tax, for 2008 as 2009, the traffic congestion remained unchanged during this period. Thus, instead of Eq. (5.1), we use Eq. (5.12) to calculate the generalized cost for 2008. The subscripts 2008 and 2009 denote which year's data are used for each term in Eq. (5.12).

$$\begin{aligned}\text{Generalized Cost}_{2008} = {} & \text{Expressway Toll}_{2008} + \text{Time Cost}_{2008} \\ & + (\text{Fuel Price Inclusive Of Tax}_{2009} \times \text{Fuel Consumption}_{2008}) \\ & + \text{Other Travel Cost}_{2008}\end{aligned} \tag{5.12}$$

- Upper-Limit Scenario
  In this scenario, an increase in traffic congestion is assumed to be fully attributable to the changes in fuel prices, inclusive of tax. In this case, we assume that while the traffic volume in 2008 changed due to the same fuel prices, exclusive of tax, for 2008 as 2009, the traffic in 2008 was as congested as the traffic in 2009. Thus, instead of Eq. (5.1), we use Eq. (5.13) to calculate the generalized cost for 2008.

$$\begin{aligned}\text{Generalized Cost}_{2008} = {} & \text{Expressway Toll}_{2008} + \text{Time Cost}_{2009} \\ & + (\text{Fuel Price Inclusive of Tax}_{2009} \times \text{Fuel Consumption}_{2009}) \\ & + \text{Other Travel Cost}_{2009}\end{aligned} \tag{5.13}$$

We calculate the generalized cost for both scenarios using Eqs. (5.12) and (5.13). The results are shown in Table 5.11.

The movement from A (the median) to D includes the effect of a sharp fall in fuel prices. Points A and B reflect the effect of the change in fuel prices under our two scenarios. Point B represents the upper-limit scenario, where it is assumed that lower fuel prices resulted in an increase in traffic congestion as in 2009. Point C illustrates the lower-limit scenario, where we assume that the lower fuel prices did not intensify traffic congestion. If we use the general equilibrium demand curve (5.11), we can estimate the traffic volumes under the upper-limit scenario (point B) and lower-limit scenario (point C), which are 44,857 and 39,313, respectively.

If we conduct the same estimation for all travel distance categories, we obtain estimated traffic volumes under both scenarios. Table 5.12 presents the results.

Table 5.11 Upper- and lower-limit scenarios of generalized cost (Tomei Expressway, Golden Week)

Category	Traveling distance (km)	2008 lower limit	2008 upper limit
		Generalized cost per car (yen)	Generalized cost per car (yen)
1	25	1,997.1	2,080.2
2	75	5,691.2	5,940.6
3	125	9,235.3	9,650.9
4	175	12,579.4	13,161.3
5	250	17,570.5	18,401.9
6	350	24,158.7	25,322.7
7	400	27,452.9	28,783.0

Table 5.12 Number of vehicles under upper- and lower-limit scenarios (Tomei Expressway, Golden Week)

Category	Traveling distance (km)	2008: lower limit	2008: upper limit
		Traffic volume	Traffic volume
1	25	592,536	615,781
2	75	191,111	201,020
3	125	44,857	39,313
4	175	52,434	44,531
5	250	77,732	64,293
6	350	93,076	81,760
7	400	7,380	5,460
	Total	1,059,126	1,052,158

5.5.2 *Social Cost*

The social cost is defined in Eq. (5.4). We calculated all of the cost items on the right-hand side of this equation in Sect. 5.4. Based on these cost items, we can obtain the social costs for 2008 and 2009. Those cost estimates are shown in Table 5.13.

Here, we refer to the social cost calculated with actual 2008 data (Table 5.13, the fourth column) as the median. The median reflects the idea that an increase in traffic congestion brought about increases in fuel consumption, time costs, and external costs. For this reason, the social cost per car is higher in 2008 than in 2009. Please note that the social cost per car is an average. Thus, it is equal to the average social cost.

In this section, we use the upper- and lower-limit scenarios to consider a decrease in fuel prices, as in Sect. 5.1. Equations (5.14) and (5.15) are used to calculate the social cost in 2008 under the lower- and upper-limit scenarios, respectively.

Table 5.13 Social cost per vehicle (Tomei Expressway, Golden Week)

Category	Traveling distance (km)	2008: lower limit (yen)	2008: median (yen)	2008: upper limit (yen)	2009 (yen)
1	25	1,459.9	1,514.1	1,542.4	1,542.4
2	75	4,379.8	4,542.2	4,627.3	4,627.3
3	125	7,299.7	7,570.3	7,712.2	7,712.2
4	175	10,219.5	10,598.4	10,797.1	10,797.1
5	250	14,599.3	15,140.6	15,424.4	15,424.4
6	350	20,439.0	21,196.9	21,594.2	21,594.2
7	400	23,358.9	24,225.0	24,679.1	24,679.1

$$\begin{aligned}\text{Social Cost}_{2008} = {} & \text{Time Cost}_{2008} \\ & + (\text{Fuel Price Exclusive of Tax}_{2009} \times \text{Fuel Consumption}_{2008}) \\ & + \text{ Other Travel Cost}_{2008} + \text{External Cost}_{2008}\end{aligned} \tag{5.14}$$

$$\begin{aligned}\text{Social Cost}_{2008} = {} & \text{Time Cost}_{2009} \\ & + (\text{Fuel Price Exclusive of Tax}_{2009} \times \text{Fuel Consumption}_{2009}) \\ & + \text{ Other Travel Cost}_{2009} + \text{External Cost}_{2009}\end{aligned} \tag{5.15}$$

The social costs calculated with Eqs. (5.14) and (5.15) are shown in the third column (lower-limit scenario) and fifth column (upper-limit scenario) in Table 5.13. The upper-limit scenario assumes that traffic in 2008 was as congested as in 2009 at the same level of fuel prices as 2009. For this scenario, the right-hand side of Eq. (5.15) takes the values for 2009 for all cost items. Therefore, the social cost for the upper-limit scenario is equal to the social cost of 2009.

As in Sect. 5.1, we take travel distance category 3 (125 km) and reproduce Fig. 5.5 for the generalized cost approach with the average social cost added. This figure corresponds to Fig. 5.4 in Sect. 5.3, in which we describe the average social cost approach.

The social costs for 2008 and 2009 are represented by the heights of points G and E, respectively. The differences in height among point G, which is the average social cost for 2008 (the median), point H, which corresponds to the lower-limit scenario, and point F are due to different assumptions concerning fuel prices and traffic congestion. As we have already mentioned, the heights of E and F are equal because the social cost for 2008 under the upper-limit scenario is identical to that for 2009.

5.5.3 Cost and Benefits

As we explained, the benefits from the new toll discount program for the Tomei Expressway service market are represented by the area below the general equilibrium demand curve. An increase in the cost under this system is viewed as an additional social cost. We take travel category 3 (a travel distance of 125 km) as an example again and present the cost and benefits in Fig. 5.6.

First, we consider the benefits. Point A denotes the generalized cost in 2008 (the median) of 9,421.7 yen, which decreases to 7,400.9 yen at Point D. The number of cars travel 125 km on the Tomei Expressway changed from 42,370 at P to 69,325 at K. The benefits from the discount program are represented by the area under the general equilibrium demand curve, and for the median case, the benefits are represented by □ADKP. Similarly, □CDKJ represents the benefits for the lower-limit scenario, and □BDKI represents the benefits for the upper-limit scenario. We can calculate the area of these rectangles because we have already obtained all of the numbers corresponding to all of the points in the figure.

For our analysis, travel distances on the Tomei Expressway are divided into seven categories, and we evaluate the benefits for each category under the median case (2008 actual data) and the upper- and lower-limit scenarios. The results are shown in Table 5.14. The sum of the benefits for all of the categories ranges from a minimum of 2,530 million yen to a maximum of 3,150 million yen. In terms of the use of the Tomei Expressway for short travel distances (categories 1 and 2), the benefits from the new discounts are relatively small compared to the longer travel distance categories, and the number of cars travelling these distances decreased from 2008 to 2009. As a result, the benefits were negative.

Next, we consider the cost. The cost incurred from the new discounts related to the Tomei Expressway is the difference between the total social cost (social cost per car multiplied by traffic volume) for 2009 minus the cost for 2008. In Fig. 5.6, point

Fig. 5.6 Generalized demand function and social cost function (Tomei Expressway, Category 3)

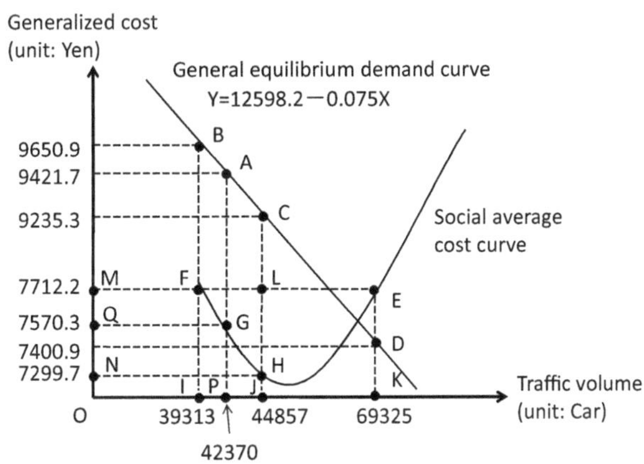

Table 5.14 Benefit for each category in the Tomei Expressway (Golden Week)

Category	Traveling distance (km)	Benefit: lower limit (million yen)	benefit: median (million yen)	benefit: upper limit (million yen)
1	25	−162.9	−183.9	−210.3
2	75	−178	−203.5	−235.6
3	125	203.5	226.7	255.9
4	175	400.7	445.8	502.4
5	250	955.6	1,062.6	1,197.3
6	350	1,096.6	1,220.5	1,376.5
7	400	210.9	234.8	264.9
	Total	2,526.4	2,803.0	3,151.1

G represents an average social cost per car of 7,570.3 yen for the median, and the traffic volume is 42,370 cars. The total social cost for category 3 in 2008, which is the year before the discount, amounted to 320 million yen (7,570.3 yen/car multiplied by 42,370 cars).

For 2009, the average social cost per car increased to E, or 7,712.2 yen, the traffic volume rose to 69,325 cars, and the social cost increased to 530 million yen (7,712.2 yen/car multiplied by 69,325 cars).

Thus, the social cost for the median in 2008 was 320 million yen, and the social cost in 2009 was 530 million yen, which means that the social cost over the year was 210 million yen (hexagon MEKPGQ). In other words, the new discount system raised the social cost for the Tomei Expressway travel distance category 3 by 210 million yen.

F and I represent the average social cost per car and traffic volume, respectively, for the upper-limit scenario, whereas H represents the average social cost and J represents the traffic volume for the lower-limit case. Therefore, the social cost that resulted from the new discount program is represented by □FEKI (the upper-limit scenario) and hexagon MEKJHN (the lower-limit scenario).

As we evaluated the benefits, we calculate the social cost for all of the rest, and the results are shown in Table 5.15. The total social cost for all of the categories ranges from 2,860 million yen (minimum) to 3,080 million yen (maximum).

The net benefits (Table 5.16), or benefits minus costs, is −80 million yen for the median and −330 million yen for the lower-limit case. The net benefits are a positive 70 million yen for the upper-limit case.

Thus, the change in the social welfare for the Tomei Expressway during the Golden Week period in 2009, compared to the same period in the previous year, is estimated to be slightly less than 100 million yen (net benefit) under the most optimistic scenario and negative (net cost) for the median and lower-limit scenarios.

Table 5.15 Social cost for the Tomei Expressway (Golden Week)

Category	Traveling distance (km)	Cost: lower limit (million yen)	Cost: median (million yen)	Cost: upper limit (million yen)
1	25	−87.8	−135.6	−172.5
2	75	−109.1	−160.3	−202.2
3	125	207.2	213.9	231.5
4	175	414.2	431.9	469.2
5	250	1,003.8	1,053.0	1,146.9
6	350	1,206.8	1,243.9	1,343.7
7	400	222.4	236.9	260.1
	Total	2,857.6	2,883.6	3,076.6

Table 5.16 Net benefit for the Tomei Expressway (Golden Week)

Category	Traveling distance (km)	Net benefit: lower limit (million yen)	Net benefit: median (million yen)	Net benefit: upper limit (million yen)
1	25	−75.1	−48.3	−37.8
2	75	−68.9	−43.2	−33.4
3	125	−3.7	12.8	24.4
4	175	−13.5	13.9	33.2
5	250	−48.2	9.7	50.4
6	350	−110.2	−23.3	32.9
7	400	−11.6	−2.1	4.8
	Total	−331.2	−80.6	74.5

5.6 Aggregate Cost and Benefits of the 1,000-Yen Expressway Discount System

5.6.1 Impact on the Tokaido Shinkansen

The sharp decrease in the expressway tolls had a substantial impact on alternative transportation systems. Because of the less expensive tolls, some of those who had used other transportation systems turned to the expressways. In fact, a decrease in the tolls to cross the Great Seto Bridge reportedly caused a substantial decrease in the number of passengers who used the ferry boats connecting the same places as the bridge. It was also reported that the number of passengers on the Shinkansen and Expressway buses decreased remarkably as a result of the discounts applied to the Tomei Expressway tolls compared with for the number of passengers in 2008. Thus, if we analyze the costs and benefits of the increasing use of expressways, we must pay adequate attention to its impact on alternative transportation systems.

The Tokaido Shinkansen train runs mostly parallel to the Tomei Expressway. It is natural that some individuals who used to take the Tokaido Shinkansen started driving on the Tomei Expressway because of the new discounts. Consequently, economic welfare decreased in the Shinkansen markets as the number of Tokaido Shinkansen passengers fell. In our study, we take the Tokaido Shinkansen as an example of an alternative transportation system to the Tomei Expressway.

The annual earnings from the Tokaido Shinkansen (Tokyo to Shinosaka) account for 86.1 % of JR Tokai's total revenue.[7] According to JR Tokai's unconsolidated balance sheet for the 2008 fiscal year, net profits amounted to 153,953 million yen.[8] Suppose the percentage of the total net income accounted for by the Tokaido Shinkansen business is equal to its share of the total earnings. JR Tokai's net income from the Tokaido Shinkansen business is then estimated to be approximately 132,554 million yen (0.86 multiplied by 153,953 million yen) for fiscal year 2008.

JR Tokai (2008) reported that 151 million people used the Tokaido Shinkansen (Tokyo to Shinosaka) in 2008. If we divide the net income from the Tokaido Shinkansen by the number of passengers, we obtain a net income per passenger of approximately 877.8 yen.

The numbers of passengers using the Tokaido Shinkansen (Tokyo to Shinosaka) during the Golden Week period in 2008 and 2009 were reportedly 3,582 thousand and 3,331 thousand, respectively.[9] In our analysis, we assume that only the expressway discount system, among other factors that affect the Tokaido Shinkansen, changed from 2008 to 2009.[10] Thus, a decrease of 251,000 passengers is assumed to have resulted from the new discount.

Using these data, we look at the general equilibrium demand curve and social cost curve for the Tokaido Shinkansen. For simplicity, we disregard external costs, such as the emissions of CO_2 and air pollutants associated with the use of the Shinkansen, and presume a horizontal social marginal curve.

We assume that the generalized cost for the Tokaido Shinkansen passengers consists of fares for passenger and express tickets and the opportunity cost of travel time, as shown in Eq. (5.16). For the case of the Tokaido Shinkansen, the area under the general equilibrium demand curve represents the benefits for the case of the Tomei Expressway. The area under the curve also represents the cost because the Shinkansen lost some of its passengers to the Tomei Expressway as a result of the new expressway discount. In Fig. 5.7, □BAEF represents the cost. The general equilibrium demand curve is horizontal, as clearly shown in Fig. 5.7, because the

[7] Cited from the following JR Tokai webpage http://company.jr-central.co.jp/company/about/area.html#main (Last access date: Oct 15, 2009).

[8] Cited from the following JR Tokai webpage http://company.jr-central.co.jp/company/achievement/finance/highlights.html (Last access date: Oct 25, 2009).

[9] Cite from the following JR Tokai webpage http://jr-central.co.jp/news/release/_pdf/000004949.pdf (Last access date: Oct 25, 2009).

[10] As mentioned in Sect. 5.3, we consider changes in fuel prices.

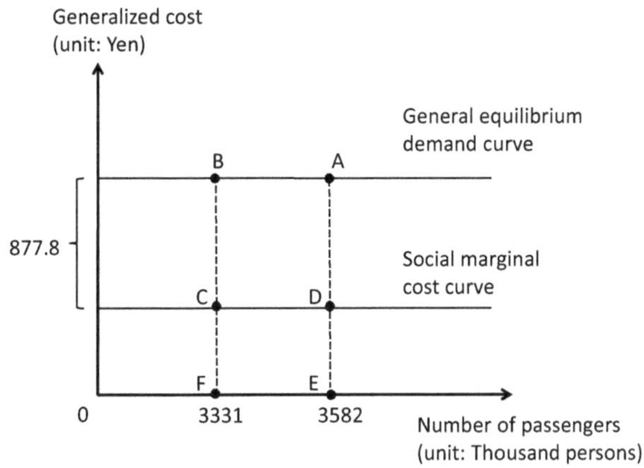

Fig. 5.7 Generalized demand function and social cost function (Tokaido Shinkansen)

Shinkansen fare and time cost are constant, regardless of whether the new discounts are introduced.

$$\text{Generalized cost} = \text{Shinkansen fare (passenger ticket and express fare)} + \text{time cost} \quad (5.16)$$

So far, we have used the average social cost to calculate the social cost. Here, we adopt the marginal cost approach to estimate the social cost for the Tokaido Shinkansen. If we suppose there is no external cost, the social marginal cost for JR Tokai is then the sum of the private marginal cost and the time cost, as expressed in Eq. (5.17). A decrease in the number of passengers decreases the area under the social marginal cost curve, which represents the total social cost, and this decrease in the area represents a decrease in the total social cost. Hence, the area under the social marginal cost curve represents the benefits from the new expressway discounts. In Fig. 5.7, □CDEF represents the benefits resulting from a decrease in the number of passengers.

$$\text{Social Marginal Cost} = \text{Private Marginal Cost} + \text{Time Cost} \quad (5.17)$$

Excluding the overlapping portion of the cost and benefits, we can represent the net benefits (negative value) for the Tokaido Shinkansen resulting from the new expressway discounts as □BADC. The net benefits are negative because the cost exceeds the benefits.

The difference between the generalized cost and social marginal cost is equal to the Shinkansen fare minus the private marginal cost. This value is the profit per passenger on the Tokaido Shinkansen, which has been already estimated at 877.8 yen. That is, the distance between the general equilibrium demand curve and social marginal curve, represented as the line segment BC (or AD), corresponds to 877.8 yen.

Combining the number of passengers in 2008 and 2009 with the net profits per capita, we can calculate a negative net benefit of −2,200 million yen (=988.8 yen/capita multiplied by the change in the number of passengers, i.e., 3,331 thousand passengers – 3,582 thousand passengers).

However, care must be taken in interpreting these results because the estimated cost (negative benefits) is the one for the route interval between Tokyo and Shinosaka, which is the route for the Tokaido Shinkansen. If we assume that passengers are uniformly distributed with respect to travel distance, the number of passengers between Tokyo and Nagoya, as a proportion of the number of passengers between Tokyo and Shinosaka, can be estimated by the working kilometers between Tokyo and Nagoya divided by the working kilometers between Tokyo and Shinosaka. The former is 366 km and the latter is 552.6 km, so the proportion is 0.662. Therefore, the cost (negative net benefits) for the route interval between Tokyo and Nagoya, as an alternative to the Tomei Expressway, is estimated at 150 million yen (=220 million yen multiplied by 0.662).

5.6.2 Net Benefits from the New Expressway Discounts

Our analysis focuses on both the Tomei Expressway and Tokaido Shinkansen, and subtracting the cost from the benefits yields the net benefits from the new expressway discount. The benefits for the Tomei Expressway are presented in Table 5.16, and the benefits for the Tokaido Shinkansen amount to −150 million yen. Thus, the net benefits for society from the new expressway discount during the Golden Week period ranges from −477 million yen (minimum) to −71 million yen (Table 5.17).

Our economic welfare analysis indicates that negative net benefits resulted from the new expressway discount during the Golden Week period with respect to the Tomei Expressway. The new expressway discount produced more costs than benefits. The result would remain unchanged even if we assume that this new holiday discount is not responsible for the increase in traffic congestion. Hence, we can conclude that the new expressway discount is not justifiable as an economic policy, as the government had initially claimed.

Regarding items other than net benefits, Tables 5.18 and 5.19 present the changes in CO_2 emissions and fuel tax revenues, respectively. Table 5.18 illustrates that the CO_2 emissions increased by 5,530 tons to 6,434 tons for the Tomei Expressway during the Golden Week period of 2009. The new expressway discount increased the traffic volume and decreased the average driving speed as a result of the heavier traffic congestion. These two effects led to an increase in CO_2 emissions. Thus, the new expressway discount worsened environmental problems.

The fuel tax revenue actually increased by 202 million yen from 2008 to 2009. The new expressway discount accounted for approximately 126–147 million yen of this total increase.

Table 5.17 Net benefit of the 1,000 Yen Expressway Discount (Golden Week)

Net benefit: lower limit (million yen)	Net benefit: median (million yen)	Net benefit: upper limit (million yen)
−477.1	−226.5	−71.4

Table 5.18 CO_2 emissions (Tomei Expressway, Golden Week)

Category	Traveling distance (km)	2008: lower limit (ton)	2008: median (ton)	2008: upper limit (ton)	2009 (ton)
1	25	1,875.6	1,909.0	1,986.0	1,625.0
2	75	1,814.8	1,857.0	1,945.0	1,522.0
3	125	710.0	671.0	634.0	1,118.0
4	175	1,161.8	1,083.0	1,005.4	1,987.0
5	250	2,460.6	2,270.0	2,073.6	4,472.0
6	350	4,124.8	3,900.0	3,691.7	6,501.0
7	400	373.8	330.0	281.7	826.0
	Total	12,521.4	12,019.0	11,617.5	18,051.0

Table 5.19 Gasoline tax revenue (Tomei Expressway, Golden Week)

Category	Traveling distance (km)	2008: lower limit (million yen)	2008: median (million yen)	2008: upper limit (million yen)	2009 (million yen)
1	25	42.8	33.4	45.3	37.1
2	75	41.4	32.5	44.4	34.7
3	125	16.2	11.7	14.5	25.5
4	175	26.5	18.9	22.9	45.3
5	250	56.1	39.7	47.3	102.0
6	350	94.1	68.2	84.2	148.3
7	400	8.5	5.8	6.4	18.8
	Total	285.6	210.2	265.0	411.7

5.7 Conclusions

This chapter focused on the Tomei Expressway during the Golden Week period and presented an economic welfare analysis of the new holiday expressway special discount program, the so-called 1,000-Yen Expressway Discount. For this purpose, we also considered the Tokaido Shinkansen as an alternative transportation system to the Tomei Expressway.

Due to a substantial decrease in toll rates, driven by the new discounts in 2009, the number of users of the Tomei Expressway increased by approximately 6 %. The decrease in tolls was beneficial for the consumers and increased their benefits. However, an increase in the social cost was caused by increased CO_2 and air pollutant emissions and the increased number of traffic accidents that resulted from more frequent occurrences of traffic congestion.

Our calculations indicate that the change in the net social surplus was negative. The new expressway discount increased the social cost by 71–477 million yen. In other words, the increases in cost due to time loss, climate change, air pollution, and traffic accidents exceeded the consumer benefits from the society's perspective. For the Tomei Expressway, the special holiday discount program was not a beneficial economic policy.

These findings have significant policy implications. First, the application of the universal discount to all expressways throughout the country makes traffic congestion heavier in expressways such as the Tomei Expressway, on which traffic congestion occurs frequently. This increased congestion reduces social welfare. Hence, it would be desirable to set appropriate discount rates for each expressway in accordance with their conditions based on some objective criteria, such as economic efficiency.

Second, an environment tax should be introduced. The topics of expressway toll discounts and toll-free expressways have been discussed in relation to user fee principles or toll-free principles after the redemption of the expressway construction cost. How to address the problem of external costs, such as CO_2 emissions and air pollution, associated with the use of automobiles has not been considered. Even if toll-free expressways were implemented and did not cause any traffic congestion, driving the expressways would still produce CO_2 emissions and air pollutants because car engines burn fuel. Apart from the discussion of toll-free expressways, taxes should be imposed on car drivers to compensate for the environmental damage resulting from automobiles based on the polluter-pays principle.

Finally, some cautions may be needed in interpreting our results First, the scope of the alternative transportation systems considered is limited. Our analysis considered only the Tokaido Shinkansen as an alternative to the Tomei Expressway. However, there are several other alternative transportation systems, such as conventional train lines, expressway buses, and local roads. Our study does not consider those systems because of the lack of data. Net benefits for conventional train lines and expressway buses are considered negative, but it is difficult to estimate whether the net benefits for local roads are positive or negative.

Second, we assume that the factors in favor of vitalizing local economies, such as an increase in tourist spending, are offset by a decrease in other consumer spending. For example, suppose that a Tokyo resident stays at home during the Golden Week period unless the new discount is applied and that he or she goes out and spends money in tourist spots if the new discount is applied. In this case, she just spends some amount of money in the tourist spots instead of spending the same amount in Tokyo. In other words, our assumption implies that the new expressway toll discount causes full income transfer from one place to another. Of course, if a tourist spends more in tourist spots than in Tokyo, the margin contributes to economic activity. We do not take this difference into account because it is difficult to quantify. These factors should also be investigated in the design of expressway toll systems in the future.

Appendix 5.1 The Model for Toll Rates

Expressway toll rates are determined by the travel distance with diminishing charge. Expressway toll rates exclusive of tax (P) are calculated using Eq. (5.1) for travel distance (D). The first formula in (5.18) is applied for $D \leq 100$. The second and third formulas are for $100 < D \leq 200$ and $200 < D$, respectively.

$$P_i = \begin{cases} 150 + 24.6 \times D \\ 150 + 24.6 \times 100 + 24.6 \times (D - 100) \times 0.75 \\ 150 + 24.6 \times 100 + 24.6 \times 100 \times 0.75 + 24.6 \times (D - 200) \times 0.7 \end{cases} \tag{5.18}$$

Actual rates are then obtained by adding the consumption tax to P. Tolls are charged in increments of 50 yen and are rounded. The tolls per car for the seven travel distance categories of 25, 75, 125, 250, 350, and 400 km are calculated using (5.18). The results are shown in Table 5.2.

After the introduction of the special holiday expressway toll discounts, a 50 % discount was applied with a cap at 1,000 yen. The fourth column of Table 5.2 presents the toll rates in 2009. Tolls for travel more than 75 km were reduced to 1,000 yen. For the category 1 travel distance of 25 km, the toll did not exceed 1,000 yen and fell to 400 yen with a 50 % discount.

Appendix 5.2 Calculation of Traffic Volumes

Assuming that the driving distances in 2008 and 2009 were identical, both in the second half of the Golden Week (May 2 to May 6) and during the entire Golden Week period (April 25 to May 6), and that the demand for travel to ICs, excluding Nagoya IC, is uniformly distributed, we can calculate traffic volumes by travel distance in Table 5.4 using the percentage of the number of cars by travel distance in the fourth and fifth columns of Table 5.3. For example, we assume that the number of cars travelling 25 km in 2008 accounted for 57 % (or 50,247 cars per day) of the total (88,000 cars per day). We also assume that all cars were standard-sized car with ETC because the breakdown of automobile types is not available.

Appendix 5.3 The Model for Traffic Congestion

The Golden Week is 12 days (17,280 min) long. We assume that traffic congestion is uniformly distributed regardless of time and place. We can then estimate the average probability of being stuck in traffic at any point on the Tomei Expressway. This probability was 0.86 (=14,872/17,280) for 2008 and 1.88 (=32,556/17,280) for 2009. These probabilities mean that the probability of an average driver being stuck in a 21.48-km-long traffic jam was approximately 86 % (or approximately 43 % for a 42.96-km-long traffic jam) on the Tomei Express way in 2008. The probability of an average driver being stuck in a 20.84-km-long traffic jam was approximately 188 % (or approximately 94 % for a 41.68-km-long traffic jam) in 2009.

It is possible that one driver driving 100 km and another driver driving 400 km face different probabilities of being stuck in traffic. Intuitively, the latter should be more likely to be stuck in traffic. We use Eq. (5.19) to perform a calibration of the estimate of the probability of being stuck in traffic for each travel distance category in the year corresponding to the categories in Table 5.2. The left-hand side of (5.19) is the probability calculated above.

$$\begin{aligned}
&(\text{Average probability of being stuck in the traffic})_t = \\
&\quad \Sigma \text{i}\{(\text{probability of being stuck in the traffic})_{it} \times \\
&\quad (\text{contribution to traffic congestion number of cars})_{it}\} \\
&\quad \text{s.t.} \quad (\text{contribution to traffic congestion number of cars})_{it} = \\
&\qquad (\text{number of cars})_{it} / (\text{total number of cars})_t \\
&(\text{Probability of being stuck in traffic})_{it} = \\
&\quad (\text{probability of being stuck in traffic})_{1t} \times (\text{distance})_i / 25
\end{aligned} \tag{5.19}$$

The probability of being stuck in traffic for the 125-km travel distance category (category 3) is set to be five-fold higher than the corresponding probability for the 25-km travel distance category (category 1). Category i's contribution (number of cars) is the share of the total accounted for by cars falling into that category.

We assume again that the consumer demand for the use of the Tomei Expressway is proportional to the travel distance. For instance, the distances between the Tokyo IC and Yokohama Machida IC and between the Gotenba IC and Numazu IC are the same (20 km), as are the traffic volumes. In addition, it is assumed that the probability of occurrence of traffic congestion is the same at any point between the Tokyo IC and Nagoya IC along the Tomei Expressway. Although these assumptions are strong, we use them to simplify our calculations.

Based on the probability of being stuck in traffic obtained by calibration, we calculate the expected queue length in traffic congestion using Eq. (5.20).

$$\begin{aligned}
&(\text{Expected queue length})_{it} - (\text{probability of being stuck in traffic})_{it} \\
&\quad \times \ (\text{average queue length in traffic congestion})_t
\end{aligned} \tag{5.20}$$

Table 5.20 summarizes the estimated probabilities of being stuck in traffic and the expected queue length in traffic congestion for each travel distance category. For

Table 5.20 Expected queue length (Tomei Expressway, Golden Week)

Category	Traveling distance (km)	Probability of being stuck in the traffic 2008	2009	Expected queue length (km) 2008	2009
1	25	0.24	0.38	5.14	7.84
2	75	0.72	1.13	15.43	23.51
3	125	1.20	1.88	25.71	39.19
4	175	1.68	2.63	36.00	54.87
5	250	2.39	3.76	51.42	78.38
6	350	3.35	5.27	71.99	109.73
7	400	3.83	6.02	82.28	125.41

example, a car in category 3, which corresponds to a travel distance of 125 km, faced, on average, a 25.71-km-long queue in traffic congestion in 2008 and a 39.19-km-long queue in 2009.

Appendix 5.4 Time Cost Estimation

Based on the expected queue length calculated by travel distance category (Table 5.20), a driver who traveled 400 km on the Tomei Expressway in 2008 would have driven 82.28 km at a slow speed due to traffic congestion and cruised another 317.7 km at a normal speed. We set the travel speed in traffic congestion to 40 km/h and the travel speed without traffic congestion to 80 km, in line with the definition of traffic congestion. We can then calculate the average travel time by travel distance category using (5.21).

$$\begin{aligned}(\text{Travel time}) = \{&(\text{Expected Queue Length})/(\text{Speed In Traffic Congestion}) \\ &+ (\text{Distance} - \text{Expected Queue Length})/(\text{Normal Speed})\}\end{aligned} \tag{5.21}$$

Table 5.5 presents the calculated travel times. It took approximately 113 min to travel 125 km in 2008, whereas the travel time was approximately 10 min greater, or approximately 123 min, for the same distance in 2009. Clearly, the reason for this increase in travel time is the increasing frequency of traffic congestion.

According to MLIT (2008a), the time cost for a passenger car averages 40.1 yen/min. We multiply the travel times shown in Table 5.5 by 40.1 yen/min to obtain the time cost per car by travel distance category. Our estimates of the time costs are shown in Table 5.6.

$$(\text{Time Cost : yen}) = 40.1(\text{yen/min}) \times (\text{travel time : min}) \tag{5.22}$$

It took 10 min longer in 2009 to travel 125 km than in 2008. Thus, the increase in the time cost was approximately 400 yen (10 min multiplied by 40.1 yen/min), which corresponds to the difference between the time cost in 2008 and that in 2009 for category 3, as shown in Table 5.6.

Appendix 5.5 Fuel Cost Exclusive of Tax and Fuel Tax

A fuel tax of 53.8 yen/litter is imposed on gasoline. The provisional tax is 25.1 yen/litter, and the remainder is based on the main rules. Hence, spending on fuel has two components: fuel cost exclusive of tax and fuel tax.

The provisional tax of 25.1 yen/litter was lifted from March 31 to April 30 in 2008 and was reintroduced on May 1 in 2008. A sharp decrease in the average gasoline price of 25.1 yen/litter occurred during the Golden Week period in 2008.

The Tomei Expressway runs through four urban areas: Tokyo, Kanagawa, Shizuoka, and Aichi. Table 5.21 presents the retail price of regular gasoline (inclusive of tax) in these areas before and after the Golden Week periods in 2008 and 2009.

Although a price differential is typically observed between Tokyo and Aichi, it is not substantial and was thus ignored in our calculations. The price on April 28 in 2008 (column 2) was very different from the price on May 7 (column 3) in the same year. As we have described earlier, the main reason for this difference was the provisional tax.

For our analysis purposes, we take the average gasoline price at the beginning (April 27 and 28) and end (May 7) of the Golden Week. Therefore, the fuel tax per liter for 2008 is assumed to be 41.3 yen (12.6 yen, half of the provisional tax rate, plus 28.7 yen according to the main rules). As a result, the gasoline price was 145.1 yen/litter (including a tax of 41.3 yen/litter) in 2008 and 117.3 yen/litter (including a tax of 53.8 yen/litter) in 2009.

Using the values calculated for fuel consumption per car (Table 5.7) and these gasoline prices, we calculate the fuel cost exclusive of tax and the fuel tax (Table 5.22).

Table 5.21 Retail price of gasoline (including tax)

Area	04/28/08 (yen/litter)	05/07/08 (yen/litter)	04/27/09 (yen/litter)	05/07/09 (yen/litter)
Tokyo	132.7	160.6	117.5	117.9
Kanagawa	129.4	158.4	117.2	116.9
Sizuoka	131.0	160.1	117.2	119.0
Aichi	129.8	158.9	116.5	116.4
Average	130.7	159.5	117.1	117.6

Source: The Institute of Energy Economics, Japan

Table 5.22 Fuel cost exclusive tax and fuel tax (Tomei Expressway, Golden Week)

Category	Traveling distance (km)	2008 Fuel cost per car (yen)	2009 Fuel cost per car (yen)	2008 Fuel tax per car (yen)	2009 Fuel tax per car (yen)
1	25	139.4	86.9	55.4	73.6
2	75	418.2	260.6	166.1	220.7
3	125	696.9	434.3	276.8	367.8
4	175	975.7	608.0	387.5	515.0
5	250	1,393.9	868.6	553.6	735.7
6	350	1,951.4	1,216.1	775.0	1,029.9
7	400	2,230.2	1,389.8	885.7	1,177.0

References

Boardman A, Greenberg D, Vining A, Weimer D (2010) Cost-benefit analysis, 4th edn. Pearson Series in Economics, Prentice Hall

Institute of Transportation Economics (2009) Brief summary of research conference on expressway discount (in Japanese) http://www.itej.or.jp/cp/wp-content/uploads/top/20091002_release.pdf. Accessed 15 May 2015

Kanemoto Y, Hasuike M, Fujiwara T (2006) Microeconomic modeling for policy analysis (in Japanese). Toyokeizaishinposha, Tokyo

Ministry of Land, Infrastructure, Transport and Tourism (2008a) Manual for cost benefit analysis (in Japanese). http://www.mlit.go.jp/road/ir/hyouka/plcy/kijun/bin-ekiH20_11.pdf. Accessed 15 May 2015

Ministry of Land, Infrastructure, Transport and Tourism (2008b) Formula and units of time cost and mileage cost by vehicle (in Japanese). http://www.mlit.go.jp/road/ir/ir-council/hyouka-syuhou/4pdf/s1.pdf. Accessed 15 May 2015

NEXCO Center (2009) Traffic situation in 2009 golden week (in Japanese). http://media2.c-nexco.co.jp/images/press_conference/64/3021438864e0061855340e.pdf. Accessed 15 May 2015

Tokai JR (2008) Annual report 2008 (in Japanese). Tokai JR, Nagoya

Varian HR (1992) Microeconomic analysis. W. W Norton & Company, New York

Chapter 6
The Evaluation of "Comprehensive Management Under the Act on the Rational Use of Energy" as a Measure to Combat Climate Change for the Hotel Industry

Abstract This chapter presents an empirical analysis of the "Comprehensive Management under the Act on the Rational Use of Energy" (Energy Conservation Act) as a measure to combat climate change. We focus on the hotel industry. Our study, which employs individual facility-level data for the hotel industry from fiscal 2002 to 2004, demonstrates that the effect of the Energy Conservation Act varies markedly across hotels. We confirm that a reduction in greenhouse gas (GHG) emissions from both heat and electricity generation was observed in the designated energy management hotels as a whole. During the same period, however, the energy consumption of non-designated small and medium-sized hotels increased, and as a result, the GHG emissions of the industry rose. Our results highlight the importance of involving small and medium-sized hotels to increase the effectiveness of climate change mitigation measures.

Keywords Climate Change Mitigation • Energy Conservation Act • Energy Management • Hotel Industry • Ex-Post Analysis

6.1 Introduction

When the first commitment period of the Kyoto Protocol began, various measures were implemented to reduce domestic greenhouse gas (GHG) emissions. However, the emissions increased by 6.8 % in 2005 compared with 1990 levels, and the 6 % reduction target established by the protocol seemed hardly achievable. The sources of GHG emissions can be divided into three sectors: industrial, commercial and residential, and transportation. The GHG emissions trends of these sectors have differed substantially. In the industrial sector, emissions have tended to decrease, presumably because of the Voluntary Action Plan on the Environment established by the Japan Business Federation.[1] Emissions from the transportation sector peaked in 2001. Conversely, emissions from the commercial and residential sector showed a sharp rise relative to 1990 levels. The subcommittee for Kyoto target achievement of the Ministry of the Environment identifies the potential for GHG emissions

[1] See Wakabayashi and Sugiyama (2007) or Sugino and Arimura (2011) for details.

T.H. Arimura, K. Iwata, *An Evaluation of Japanese Environmental Regulations*,
DOI 10.1007/978-94-017-9947-8_6

reductions with low marginal costs in the commercial and residential sector.[2] Thus, our focus is on the commercial and residential sectors, and we evaluate the "comprehensive management under the Act on the Rational Use of Energy" (Energy Conservation Act, hereafter), which was expected to contribute to GHG emission reductions before the first commitment period of the Kyoto Protocol began.

After the oil crisis, the "Energy Conservation Act" was introduced in 1979 with the goal of efficient energy use in the context of energy security. Among policy measures under the act, the top runner programs, including fuel efficiency standards, are well known.[3] In addition, the act attempts to promote energy savings in plants and offices. For example, the act requires regulated facilities to hire qualified energy managers. Moreover, the act also requires facilities to report their energy consumption. It also establishes a target of 1 % reduction in energy intensity.[4] The objective is to reduce energy consumption rather than GHG emissions, but reducing energy consumption results in a decline in fossil fuel consumption. Therefore, the goal of the Energy Conservation Act was quite close to the GHG emission reduction even before climate change was recognized as a major policy issue. Quite naturally, when the first commitment period began, the act was viewed as a measure to combat climate change and to contribute to achieving the Kyoto Protocol target plan published by the Cabinet Public Relations Office (2005). The plan emphasized that the objective of the Energy Conservation Act included reducing GHG emissions. The plan projected a reduction in emissions in the commercial and residential sector by 3 million tons.[5]

As discussed above, the Energy Conservation Act is considered a strategy to mitigate GHG emissions and address climate change. Nonetheless, few studies have been conducted to examine the quantitative effects of the act with respect to reducing CO_2 emissions because of the limited availability of relevant data. In addition, the progress toward achieving the target of a 1 % reduction in energy intensity has not been thoroughly reviewed. Based on their literature review and interviews, Sugiyama and Tanabe (2002) note that businesses using a large amount of energy comply with the energy conservation act, whereas businesses consuming a small amount of energy do not. However, Sugiyama and Tanabe (2002) do not address the quantitative effects of the act in terms of reducing energy consumption. Ito and Terao (2005) investigate the determinants of fuel consumption in the paper and pulp industry, and they confirm that the quantity produced has a substantial effect on fuel consumption.

[2] Available at http://www.env.go.jp/council/06earth/r062-01/index.html.

[3] The energy conservation act has been amended several times.

[4] The energy conservation act defines energy intensity as energy consumption divided by variables such as floor space, which is closely related to energy consumption.

[5] See the Kyoto Protocol target achievement plan attachment to Table 1 "List of measures and policies related to CO_2 emissions from the use of energy," which is available on the website above.

Although the energy conservation act is applied to all industries, we focus on a specific industry to consider its effects. We select the hotel industry[6] for the following reasons. First, the act was relatively recently applied (in 1999) to the hotel industry. Thus, the effect of the act, if any, can be easily identified. Second, as we discuss below, there are Type 1 and Type 2 designated facility operators in the industry, which allows us to explore the difference in the effects of the act on the two types of facilities. Third, the hotel industry belongs to the commercial and residential sector, in which further reductions in CO_2 emissions are required.

The next section describes the energy conservation act, its assessment and its coverage. Section 6.3 introduces the analytical methodology and presents the results and a discussion. Section 6.4 concludes the chapter.

6.2 The Act on the Rational Use of Energy (Energy Conservation Act)

6.2.1 Overview

The Energy Conservation Act has been amended several times. We use data from 2002 to 2004. For this reason, we explain the changes of the Energy Conservation Act during that period.

The act categorizes energy into two types: heat and electricity. "Heat" refers to crude oil; volatile oil; naphtha; heating oil; light oil; bunker A, B, and C; petroleum asphalt; petroleum coke; petroleum gas; combustible natural gas (e.g., LNG); coal; coal coke; coal tar; coke-oven gas; blast furnace gas; and converter gas. Depending on the volume of heat and electricity consumption, facilities are subject to different types of regulations under the act.

The major form of regulation was the designation of *energy management facilities*. Once facilities are designated as energy management facilities, facility managers become subject to regulatory oversight. By 1999, the manufacturing, mining, electric power supply, gas supply and heat supply industries were designated as energy management facilities. After the amendment of the act in 1999, its coverage was expanded to include the hotel industry.

Under the 1999 amendment, the facilities covered under the previous act were categorized as "Type 1 energy management facilities" (Type 1). Under the 1999 amendment, smaller facilities became subject to the regulation as "Type 2 energy management facilities" (Type 2). The Type 2 designation became applicable to all sectors rather than to specific manufacturing sectors.

The act was amended again in 2003. Under this amendment, larger-scale facilities among the Type 2 designated facilities were upgraded to Type 1 facilities. The

[6] The "hotel industry" also includes Japanese-style inns.

distinction between Type 1 and Type 2 designations is as follows. Regarding heat, facilities that consume more than 1,500 kl of various fossil fuels per year in crude oil equivalents are designated as *Type 2 heat management facilities*. Facilities using more than 3,000 kl per day are designated as *Type 1 heat management facilities*. For electricity, facilities that use at least 600 million KWH of electricity per year are designated as *Type 2 electricity management facilities*, and those consuming more than 1,200 KWH per year are designated as *Type 1 electricity management facilities*.

One major requirement of the Energy Conservation Act for facilities is to employ heat or electricity management managers. Once facilities are designated as heat management facilities, they must employ heat managers with certificates from the Ministry for Economy, Trade and Industry or from a related agency. Similarly, once facilities are designated as electricity management facilities, they must employ certified electricity managers. The number of managers required depends on the size of the facility, as shown in Table 6.1. These managers are expected to promote energy conservation at their facilities.

To reduce energy consumption, the act requires facilities to be engaged in "energy management" activities. First, heat/electricity managers are expected to measure and record energy consumption and to use this information to manage the use of energy. Second, the managers must check equipment and facilities. For example, regarding "heat," the managers report "the status of the establishment of management standards,[7]" "the status of compliance regarding measurement and recording," "the status of compliance regarding maintenance and repairs" and "the status of measures for new equipment" using five items: combustion equipment, heat utilization equipment, loss prevention equipment, heat recovery equipment and electric power/cogeneration equipment.[8] Similarly, regarding "electricity," the managers are expected to report their own measures for power receiving and transforming equipment/distribution equipment, electricity utilization equipment and electricity equipment (lighting) (see Sugiyama et al. 2010). Third, heat/electricity managers strongly encourage workers to engage in energy-saving practices in their daily activities. Finally, the act also requires the periodic submission of reports on the use of both heat and electricity. The reports provide information on energy consumption in these facilities and indicates whether various measures and actions to reduce energy consumption have been adopted.

Based on these "comprehensive energy management" efforts, the energy conservation act aims to achieve a 1 % reduction in energy intensity every year. However, this goal may have not been recognized seriously, and progress has not

[7] "Management standards" refers to a management manual for operation, management, measurement, recording, maintenance and repairs regarding the rational use of energy. A sample of management standards is available from the Energy Conservation Center Japan (2005).

[8] Cogeneration equipment uses waste heat from diesel engines or gas turbines for power generation to simultaneously generate both electricity and useful heat to improve energy efficiency.

Table 6.1 Number of energy managers required by act

1st class designated mining, electricity/gas/heat supply factories	
Annual fuel consumption	Number required
3,000 or less than 100,000 kl-oe	1
100,000 kl-oe or more	2
1st class heat designated manufacturing factories	
Annual fuel consumption	Number required
3,000 or less than 20,000 kl-oe	1
20,000 or less than 50,000 kl-oe	2
50,000 or less than 100,000 kl-oe	3
100,000 kl-oe or more	4
1st class electricity designated manufacturing factories	
Annual electricity consumption	Number required
12,000 or less than 200,000 MWh	1
200,000 or less than 500,000 MWh	2
500,000 MWh or more	3

Source: *Japan Energy Conservation Handbook 2003/2004* (ECCJ) http://www.asiaeec-col.eccj.or.jp/databook/2003-2004e/04_03_01.html

been assessed. Therefore, the target was not sufficiently reviewed; hence, the Kyoto Protocol target achievement plan did not include the emission intensity target.

Here, we briefly explain the difference between the Type 1 and Type 2 designations. First, these designations differ with respect to obligations. Type 1 facilities are required to develop and submit medium- and long-term energy savings plans, whereas Type 2 facilities are not.

The penalties also differ between Type 1 and Type 2 facilities. There are several types of penalties under the act, including *advice* from the regulator, *on-site inspections* and *public announcements*. *Advice* is the weakest penalty, whereas *public announcement* is considered the most severe. If facilities do not make progress in reducing energy consumption, then Type 2 facilities receive *advice* from the regulator. By contrast, Type 1 facilities must develop and implement plans to improve the rational use of energy. Hence, Type 1 facilities confront stricter energy consumption guidelines than Type 2 facilities and have stronger incentives to use less energy.

How is "comprehensive energy management" enforced under the energy conservation act? The regulatory authorities conduct on-site inspections to ensure that facilities comply with the act. Facilities that are subject to on-site investigation are required to self-inspect their facilities primarily with respect to the establishment of management standards and the status of measurement and recording, maintenance and repairs and to rank each item using three grades (ranging from zero to two points). If the overall score for a facility is below 50, then the facility is recognized as a "facility that substantially lacks in efforts to rationalize energy consumption" and is listed for on-site inspection and required to develop energy-saving plans (ECCJ 2005).

How often do facilities receive penalties? Let us examine the results from 2005 to 2009. A total of 1,759 facilities received *advice*,[9] the weakest penalty from the regulator. On-site inspections were conducted at 56 facilities. However, there were no *public announcements* of the names of facilities that exhibited poor performance. Overall, although numerous penalties were imposed, the most stringent penalty was not used.

6.2.2 Assessment Standards of the Energy Conservation Act

The Kyoto Protocol target achievement plan projected that the industrial sector and commercial and residential sector would reduce CO_2 emissions by 170 and 300 million tons, respectively, as a result of the comprehensive energy management provisions of the energy conservation act. The Kyoto Protocol target achievement plan (p. 150, 2006) developed by the climate change prevention headquarters stipulates countermeasures and policies to meet the target related to the "improvement of energy efficiency in office buildings" as follows:

- Improved energy intensity of facilities that are newly designated as Type 2 facilities
- Improved energy intensity of facilities that are newly upgraded from Type 2 to Type 1 facilities

That is, the Kyoto Protocol target achievement plan aims to promote improved energy intensity through "comprehensive energy management" under the Energy Conservation Act and to reduce CO_2 emissions in the process. Hence, we evaluate the Energy Conservation Act by investigating the following two points:

- Whether energy intensity has improved (improvement in energy intensity)
- Whether energy consumption has decreased (lower energy consumption)

6.2.3 Evaluation Target

Given the purpose of the Kyoto Protocol target achievement plan, it would be appropriate to evaluate the effect of the Energy Conservation Act in terms of the aspects described below:

- The effect of newly designating facilities as Type 2 (the amendment of the act in 1999)
- The effect of upgrading facilities from Type 2 to Type 1 (the amendment of the act in 2003)

[9] *Advice* from regulators is considered a penalty in Japanese society.

For the first effect, one means of assessing the effects of a new energy management facility designation is to compare the fuel consumption of a facility before and after the designation. However, this approach is impossible because of the lack of data. Another strategy is to compare designated energy management facilities with non-designated facilities. Unfortunately, data on non-designated facilities are not available.[10] Therefore, we primarily use descriptive statistics to analyze the first effect, and we conduct a detailed econometric analysis to assess the second effect.

We also analyze the efforts of individual facilities to reduce energy consumption. On-site inspection, as a measure to ensure facilities' strict compliance with the Energy Conservation Act, is conducted based on whether a facility exhibits a lack of effort to rationalize energy use. To evaluate effort levels, commitments by individual facilities, such as the establishment of management standards, are assessed. We will evaluate whether the commitments of individual facilities are effective. Specifically, we investigate whether, energy-saving activities regarding electricity and heat are effective in reducing energy consumption.

6.3 Estimation Methodology and Results

6.3.1 Data and Evaluation Criteria

Once facilities are designated as Type 1 or Type 2 energy management facilities under the Energy Conservation Act, they must submit periodic reports. In these reports, facilities are required to provide information on their energy consumption levels (electricity and heat) and the amount of energy consumption per unit of production. In addition, each facility's energy-saving activities regarding electricity and heat are also included in the reports. We used these periodical reports in the evaluation.

The Energy Conservation Act has covered the hotel industry since the 1999 amendment to the act. Because of data limitations, we used data from 2002 to 2004. A total of 142 facilities[11] that submitted periodical reports from 2002 to 2004 are analyzed in this study. We hypothesize that the improvements in energy efficiency and reductions in energy consumption observed from 2002 to 2004 are the effect of the facilities becoming designated facilities under the act.

In the periodic reports, two types of energy consumption are recorded: heat and electricity. Regarding "heat," the facilities report fuel consumption (crude petroleum equivalent; kl). In terms of "electricity," the facilities report electricity

[10] Even if data were available for non-designated facilities, a comparison would be inadequate because the sizes of these firms differ from those of designated facilities under the provisions of the energy conservation act.

[11] All designated facilities (heat type 1, heat type 2, electricity type 1 and electricity type 2) are included in the sample of 142 facilities.

Table 6.2 Descriptive statistics: energy consumption

Fiscal year	Heat consumption (kl) Number of observations	Mean	Standard deviation	Electricity consumption (thousand kWh) Number of observations	Mean	Standard deviation
2002	86	3,087.80	2,042.60	117	13,257.70	9,980.00
2003	86	2,914.80	1,822.00	126	12,905.40	9,490.50
2004	87	2,988.40	1,752.90	126	13,118.80	9,836.30

consumption (kWh). Table 6.2 presents descriptive statistics on energy usage. Heat consumption decreased from 2002 to 2003 and subsequently increased from 2003 to 2004. In total, energy usage decreased by 3.2 %, from 3,088 to 2,988 kl, which is equivalent to a reduction of 22,650 t-CO_2.[12]

Even if GHGs from heat consumption have declined, total GHG emissions may not have declined if fossil fuel consumption were replaced by electricity usage. According to Table 6.2, electricity usage exhibits the same trend as heat. Electricity consumption decreased from 2002 to 2003 and subsequently increased from 2003 to 2004. These figures show a 1.0 % decline in total energy usage, from 13,258 to 13,119 kWh, which is equivalent to a decline of approximately 7,436 t-CO_2.[13]

Therefore, by becoming designated facilities under the Energy Conservation Act, these facilities were able to reduce GHG emissions by decreasing the use of heat and electricity. This finding demonstrates the effects of the Energy Conservation Act.

In this research, we also examined the improvement in energy intensity as an evaluation criterion. Energy intensity in both heat and electricity is calculated as follows:

$$(\text{Energy intensity}) = (\text{total consumption})/(\text{production volume indicator}).$$

Because the indicators of production volume vary across hotels, it is impossible to simply compare the energy intensity levels of facilities. For example, some hotels may use the number of rooms as the indicator, whereas others may consider floor space. If a hotel selects the number of rooms as a production indicator, then it may not change from 2002 to 2004. Thus, we must construct an index that is independent of the indicators of production volume. We define E^t_{ij} as the energy consumption of facility i in year t and Y^t_{ij} as a production volume index (where $j = 1$ indicates heat and $j = 2$ indicates electricity). Energy intensity can be defined as E^t_{ij}/Y^t_{ij}, and the

[12] To calculate the CO_2 emission coefficient, we used the calculation method 38.2 GJ/kl × 0.0187 t-CO_2/GJ × 44/12 from the "GHG Emissions Accounting and Reporting Manual" (MOE 2013). For the number of facilities, we used the number in 2004 in the calculation because the number of targeted facilities has changed.

[13] For the CO_2 emission coefficient, we used 0.425 kg-CO_2/kWh, which can be found in The Federation of Electric Power Companies of Japan (2006). As in the case of heat, we used the number of facilities in 2004 in the calculation.

Table 6.3 The change in energy intensity from 2002 to 2004

	Mean	Standard deviation	Max	Min	Number of observations
Heat	96.2	8.7	119.7	71.2	84
Electricity	99.4	5.0	112.1	81.9	107

rate of change in energy intensity from the previous year can be described as follows:

$$\Delta Efficiency_{ijt} = \frac{E_{ij}^{t}/Y_{ij}^{t}}{E_{ij}^{t-1}/Y_{ij}^{t-1}} \times 100$$

The units in the numerator and denominator are identical for each hotel; thus, $\Delta Efficiency_{ijt}$ represents a common index of energy efficiency improvement across hotels.

The descriptive statistics for this index are presented in Table 6.3. This table depicts the change in energy intensity from 2002 to 2004. According to the results, energy intensity in heat consumption improved by more than 3 %. The target level in the Energy Conservation Act is 1 % per annum; thus, the 2-year target is 98 ($0.99 \times 0.99 = 0.9801$). Therefore, with regard to heat, energy efficiency improved above and beyond the target level. It is worth considering the size of the variation in the change in energy intensity. In the case of heat, the change in intensity ranges from 71.2 (30 % improvement) to 119.7 (20 % decline).[14]

In the case of electricity, the change in energy intensity from 2002 to 2004 is a 0.6 % improvement. Although the target established by the Energy Conservation Act has yet to be reached, this figure shows that energy efficiency is indeed improving. The variation in the change in energy intensity is also large in the case of electricity.

According to Table 6.3, the mean is under 100 for both heat and electricity. However, it is overly simplistic to conclude that the Energy Conservation Act reduced energy intensity. Therefore, we will examine the hypothesis that the mean energy intensity relative to that of 2 years ago is statistically significantly under 100. In the case of heat, the null hypothesis was rejected at the 1 % level based on the results of a one-tailed t-test. In other words, in terms of heat, the energy intensity of the designated facilities has improved. In the case of electricity, however, the null hypothesis was not rejected. Therefore, energy intensity in electricity has not been improved.

[14] For a more detailed distribution, please see Arimura and Iwata (2007).

6.3.2 Econometric Model and Data

Next, we examine whether energy efficiency improved following the upgrade of facility designations from Type 2 to Type 1. The volume of energy consumption in hotels may be affected by several factors in addition to conditions such as whether the facilities are designated energy management facilities or whether they are designated as Type 1 or Type 2. For example, the capacity of the hotels is an important factor. In addition, electricity demands for air conditioning may increase during hot summers. Therefore, it is necessary to differentiate between factors related to the Energy Conservation Act and others that influence energy consumption. Hence, we perform the following linear regression analysis to obtain additional information on the relationships between several independent variables (variables regarding Type 1 or Type 2 designations, hotel characteristics and weather conditions) and a dependent variable (an efficiency improvement index). Using this analysis, we can examine whether the effects of those independent variables on the dependent variable are statistically significant.

We estimated two models: (1) a model for efficiency and (2) a model for the volume of consumption. In the first analysis, we examined the following econometric model using the efficiency improvement index that was defined in the previous section.

$$\Delta Efficiency_{ijt} = \theta_j + \eta_j^1 D_{ijt} + \eta_j^2 PRACTICE_{ijt} + \alpha_j^1 H_{it}^1 + \alpha_j^2 H_{it}^2 + \beta_j^1 X_{it} + \beta_j^2 YEAR_t + \varepsilon_{it} \tag{6.1}$$

where i denotes the facility number, j denotes the energy type (heat or electricity) and t refers to time.

Here, we explain our independent variables. First, we included a dummy variable D_{it} (Type 1 dummy) that takes a value of 1 when a hotel is designated as a Type 1 facility under the Energy Conservation Act and takes the value 0 otherwise. A hotel is expected to improve its efficiency level by being designated as a Type 1 facility. In addition, Type 1 facilities must comply with more stringent regulations than Type 2 facilities. Therefore, we predict that the sign of the coefficient for D_{it} will be negative.

Next, we focused on each hotel's efficiency improvement activities. In the mandated periodical reports, in addition to the volume of energy consumption, hotels are required to report energy management practices, such as the following: (i) the degree to which managerial standards have been established, (ii) the compliance level with respect to measurement and recording, (iii) the level of compliance regarding maintenance and inspection, and (iv) the measures for new establishment. In terms of "heat," these four actions are reported by each facility as follows: "combustion equipment" (heat 1), "heat utilization equipment" (heat 2), "heat utilization equipment with loss prevention" (heat 3), "equipment with exhaust heat recovery" (heat 4) and "cogeneration equipment" (heat 5). Similarly, for "electricity," four actions are reported for each facility: "heat utilization equipment" (electricity 1), "substation and/or electric power distribution equipment" (electricity 2), "electricity utilization equipment" (electricity 3) and "lighting equipment" (electricity 4). In this analysis, we focused on "the state of establishing management standards" and examined whether this action contributes to CO_2 reduction. These energy management actions are included in the estimated equation as a vector of dummy variables denoted $PRACTICE_{it}$. We expect that the sign of the coefficient on these variables will be negative.

Efficiency improvement may be influenced by various other factors. First, the capacity of a hotel is a significant factor. We included the number of restaurants, lounges and bars as well as the square of each variable in the indicator of hotel capacity H^1_{it}. Restaurants, lounges and bars are expected to be responsible for a larger share of energy consumption compared with ordinary guest rooms. The number of restaurants, lounges and bars may indicate another aspect of energy demand owing to hotel capacity. A dummy variable H^2_{it} that captures the type of hotel (comprising six types, such as a city hotel or an inn with hot spring bathing facilities) is also included. This variable may indicate the different levels of energy consumption by guests

Because air conditioning and heating represent substantial sources of energy demand, we included "the number of days with temperatures of 35 °C or higher (days at 35 °C or higher)" and "the number of days with temperatures below 0 °C (days at 0 °C or below)" in the vector denoted X_{it}.

Finally, to capture the trend, we included $YEAR_t$, a year dummy. θ, α, β and η are the parameters. In particular, η is a parameter that indicates the effect of each energy management action and that of Type 1 designation.

In the second analysis, we used the decline in energy consumption as the dependent variable. The change from the previous fiscal year (first difference[15]) is employed because it may better reflect the effect of each energy management action. The equation to be estimated is presented below.

$$\Delta E_{ijt} = \theta_j + \eta^1_j D_{ijt} + \eta^2_j PRACTICE_{ijt} + \alpha^1_j H^1_{it} + \alpha^2_j H^2_{it} + \beta^1_j X_{it} + \beta^2_j YEAR_t + \varepsilon_{it} \tag{6.2}$$

The dependent variable can be defined as follows: $\Delta E_{ijt} \equiv E^t_{ij} - E^{t-1}_{ij}$. The expectation regarding each coefficient in Eq. (6.1) also apply to those of Eq. (6.2).

Information on each energy management action (*PRACTICE*) was obtained from the periodical reports. Information on hotel capacity (H^1) and hotel types (H^2) was based on hotel websites, travel agencies or financial reports. Weather

[15] The first difference approach is a transformation of a time series constructed by taking the difference between previous year and the current year. If the difference is one year, then it is called "first difference," and if the difference is two years, then it is termed "second difference." We use first differences in this study.

information (X) was obtained from the website of the Japan Meteorological Agency.

The Type 1 dummy indicates whether the status of a facility was newly raised from Type 2 to Type 1 based on the 2003 amendment. The distinction between Type 1 and Type 2 can be obtained from the database of the Agency for Natural Resources and Energy.

We used data from 2002 to 2004 in this analysis. The dependent variables are the first difference and efficiency change (the change in energy intensity); therefore, 3 years of data are used for the estimation. Thus, we have one fewer year in this analysis.

6.3.3 Results: The Effect of Being Raised to a Type 1 Facility

Table 6.4 presents the results of the econometric analysis of energy efficiency improvements and the change in consumption. This analysis is the result of estimating Eqs. (6.1) and (6.2) in the case of heat.[16]

Model A estimated the effects of a Type 1 designation based on energy intensity. The dependent variable is the change in energy intensity for heat consumption. A negative sign on the coefficient indicates that energy intensity has decreased, and thus, efficiency has improved.

According to the results, the Type 1 dummy is negative and statistically significant at the 10 % level. This result indicates that facilities that are designated as Type 1 are likely to improve their energy intensity compared with Type 2 facilities. Based on the coefficient, the improvement is estimated at approximately 2 % per year.

Next, in the results from model B, which estimated Eq. (6.2) using the change in heat consumption as the dependent variable, the coefficient of the Type 1 dummy is also negative and significant at the 10 % level. This result is consistent with the hypothesis that Type 1 designated facilities improved their efficiency levels to a greater extent than Type 2 facilities. By being raised to Type 1, a facility is able to reduce 74.4 kl of crude oil consumption per year. This result indicates that the total effect of being raised to a Type 1 facility over the period from 2002 to 2004 is approximately 13,000 t-CO_2 reduction. Regarding energy management practices, the coefficient of "heat 4" in model B is negative and statistically significant, but the other coefficients are not significant.

In both models, the year dummy is positive and significant, which indicates that energy efficiency has declined and that the amount of consumption has increased in general. In terms of the effect of weather, the coefficient of the variable "days at 35 °C or higher" is positive and significant. This finding confirms that energy consumption increases when the temperature is high. In addition, the coefficient

[16] We conducted a White test. If the null hypothesis of homoskedasticity was rejected, then we performed a robust estimation.

Table 6.4 Estimation results: heat

	Energy intensity	Energy consumption
Variables	Model A	Model B
Type 1 dummy	−2.81*	−74.4*
Heat 1	−0.33	25.5
Heat 2	1.11	−6.38
Heat 3	2.05	88.3
Heat 4	−0.22	−80.9*
Heat 5	−3.26	−35.4
Days with 35 °C or more	0.17	8.32**
Days with 0 °C or less	0.04	1.67**
Capacity of the hotel	0.00	0.06
Capacity^2	−2.40E-07	−3.24E-06
Number of restaurants, lounges and bars	0.65**	41.2***
Number^2	−0.02**	−1.82***
Year dummy	5.38***	121.7***
Type dummy	YES	YES
Adjusted R-squared	0.27	0.65
F· Wald (P value)	4.09 (0.00)	16.1 (0.00)
White (P value)	132.9 (0.37)	132.3 (0.36)

Note: ***, ** and * indicate significance at the 1 %, 5 % and 10 % levels, respectively. The numbers of observations are 150 in Model A and 147 in Model B. The type dummy denotes the type of hotel (six types, such as a city hotel or an inn with hot spring bathing facilities). These results are taken from Arimura and Iwata (2008)

of the variable "days at 0 °C or below" becomes positive, which indicates that heat consumption increases when the temperature is low.

Table 6.5 presents the results of an econometric analysis of electricity with respect to energy efficiency improvement and change in consumption. This analysis is the result of estimating Eqs. (6.1) and (6.2) for electricity.[17]

In both models, the coefficients of the Type 1 dummy are negative. This result is consistent with the hypothesis that Type 1 designated facilities improved their efficiency to a greater extent than Type 2 facilities, although the result is not statistically significant.

In contrast to the case of the heat, none of the energy management practices has a significant effect. The year dummy is positive and significant, which is the same as in the case of heat. In contrast to the case of heat, the explanatory power of the variables is poor in the electricity estimates, and the overall effect is difficult to capture using observable independent variables.

[17] In the case of electricity, we also conducted robust estimation when the null hypothesis of homoskedasticity was rejected.

Table 6.5 Results (Models C and D)

	Electricity energy intensity	Electricity consumption
Variables	Model C	Model D
Type 1 dummy	−0.72	−58.4
Electricity 1	−0.87	−58
Electricity 2	1.01	78.8
Electricity 3	−0.59	−1.57
Electricity 4	0.42	32.6
Days with 35 °C or more (CDD)	−0.01	4.15
Days with 0 °C or less (HDD)	−0.02	−2.95
Capacity of the hotel	0.00	0.35**
Capacity^2	−2.87E-07	−0.0001187**
Number of restaurants etc.	0.2	−23.7
Number^2	0.00	1.24
Year dummy	2.73***	332.0***
Type dummy	YES	YES
Adjusted R-squared	0.05	0.11
F· Wald (P value)	2.42 (0.00)	2.52 (0.00)
White (P value)	164.9 (0.00)	191.7 (0.00)

Note: ***, ** and * indicate significance at the 1 %, 5 % and 10 % levels, respectively. The numbers of observations are 211 in Model C and 223 in Model D. The type dummy indicates the type of hotel (six types, such as a city hotel or an inn with hot spring bathing facilities)

6.3.4 Discussion

The results of the quantitative analysis are summarized below. First, the results in the previous section indicate that the consumption of heat and electricity has decreased in the designated facilities.

Second, the effect of being raised to Type 1 facilities is an approximately 2 % reduction in energy intensity for heat consumption. However, this effect was not observed in the case of electricity. In addition, this "upgrade" to Type 1 is effective in reducing heat consumption. However, we do not observe an effect of upgrading in the case of electricity consumption.

Third, some energy management actions improve energy intensity with respect to heat. However, in the case of electricity, we fail to observe such effects. The overall results regarding energy consumption and CO_2 emissions are summarized in Table 6.6.

In the following, we examine the results in terms of two aspects: (A) the scope of the regulation and (B) variation in the reduction effect.

(A) The scope of the regulation

Figure 6.1 depicts the change in total energy consumption and total energy consumption per unit of floor space for the hotel industry since 1990. These data also include the energy consumption of facilities that are not covered by the act.

Table 6.6 The reduction effect of energy management practices based on the act

	The effect of reducing energy consumption		CO_2 reduction (t-CO_2)	
	Heat	Electricity	Heat	Electricity
1. By designating facilities	Intensity 3.8 %	Intensity 0.6 %	2.3 million	0.74 million
	Consumption 3.2 %	Consumption 1.0 %		
2. By upgrading to type 1	Energy intensity 2.8 %	Unidentified	1.3 million	Unidentified
	Consumption 2.4 %			

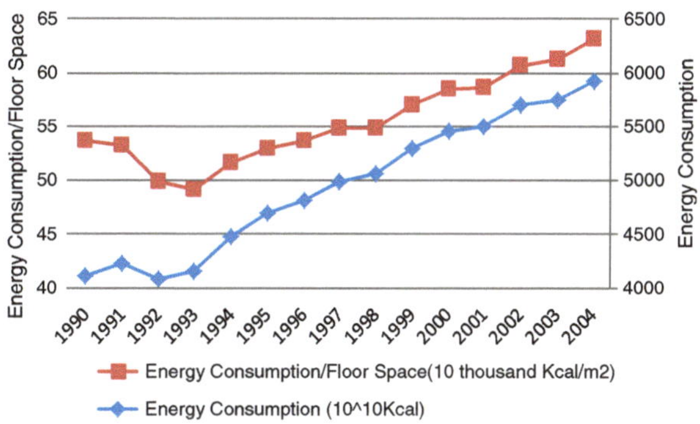

Fig. 6.1 The change in energy consumption and energy consumption per unit of floor space in the hotel industry (Data: The Energy Data and Modeling Center, the Institute of Energy Economics, Japan 2005)

From 2002 to 2004, both energy consumption and energy intensity increased in the industry as a whole. In contrast, our analysis showed that total energy consumption and intensity in the designated facilities improved from 2002 to 2004. In other words, heat and electricity consumption was reduced or controlled in the facilities that were subject to the act. Therefore, small or medium-sized facilities that are not regulated by the act exhibited increased energy consumption. This increased consumption surpassed the effect of reductions in the designated facilities.

This finding indicates an important implication for policies that are designed to produce climate change mitigation. To reduce total GHGs, we must focus on controlling the emissions of hotels that are not regulated by the act or on achieving further reductions at designated hotels.

Our empirical results show that an "upgrade" to Type 1 effectively reduced GHG emissions in the facilities under study. Based on this argument, it would be effective to expand the scope of facilities covered by the Type 1 designation to realize GHG emissions reduction in the designated facilities. However, the marginal abatement cost may already be high in the previously designated facilities as a result of various efforts. Therefore, it would be important and effective to reduce emissions from facilities that are not covered by the act. Thus, it is important to include small and medium-sized hotels in the regulation.

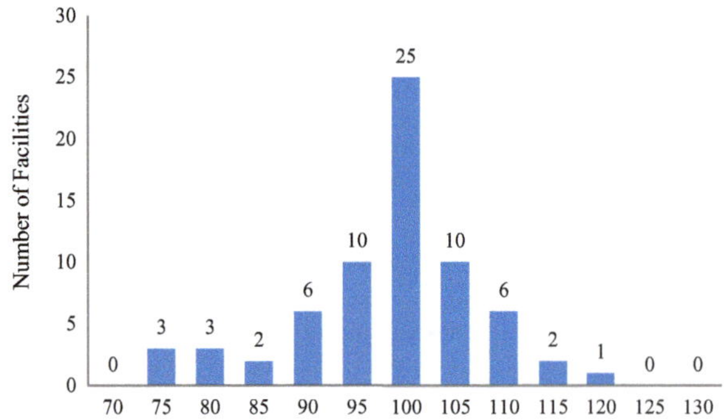

Fig. 6.2 The change in the energy intensity of heat consumption relative to 2 years ago (2002–2004). Note: The *vertical axis* indicates the number of facilities, and the *horizontal axis* indicates the change in energy intensity of heat consumption relative to 2 years ago (100 indicates no change)

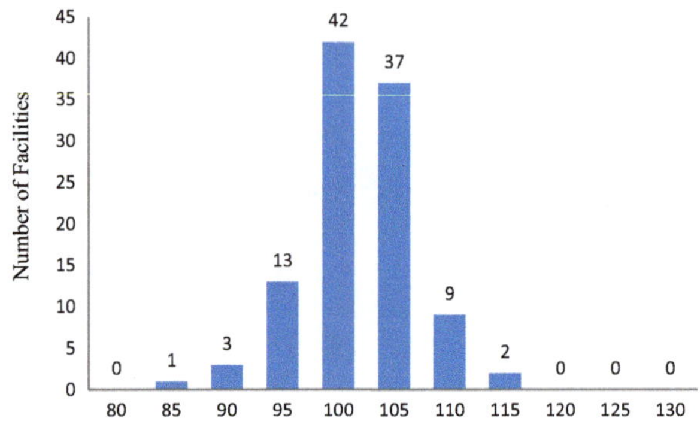

Fig. 6.3 The change in energy intensity of electricity consumption relative to 2 years ago (2002–2004). Note: The *vertical axis* indicates the number of facilities, and the *horizontal axis* indicates the change in energy intensity of electricity relative to 2 years ago (100 indicates no change)

(B) Variation in the extent of reductions

Table 6.3 clearly shows that the degree of improvement varies widely among the facilities.[18] Figure 6.2 presents a histogram depicting the change in energy intensity of heat relative to 2 years ago. Three facilities have improved their energy by 20 %, whereas the efficiency of one facility actually declined by 20 %. In addition, the change in electricity intensity is depicted in Fig. 6.3. The efficiency of one facility improved by more than 10 %, whereas two facilities experienced declines in efficiency of more than 10 %. Overall, some facilities substantially reduced their GHG emissions, but others increased their GHG emissions. This disparity in improvement poses a problem from an equity perspective. In the present situation, facilities that are not eager to improve their energy efficiency are free riding on the efforts of facilities with substantial improvements. From an equity perspective, it may be desirable to introduce enforceable instruments that may improve the energy intensity levels of facilities with poor performance to realize further emissions reductions in the hotel industry. However, it is possible that high marginal abatement costs may be associated with facilities that are not reducing their emissions. In

[18] We also conducted a "Quantile Regression," which considers the importance of unevenness. The results of this regression are presented in the Appendix 6.1.

this case, mandating emission reductions is a problem from an efficiency perspective. In principle, it is important to begin reducing emissions at facilities with low marginal abatement costs. Market-based instruments, such as emissions trading schemes and environmental taxes, are effective in this case.

6.4 Conclusions

This chapter performed a quantitative analysis of the effect of "comprehensive energy management" under the Energy Conservation Act on reductions in energy consumption and CO_2 emissions. Our analysis demonstrated that the designated facilities were able to reduce their GHG emissions by 3.2 % by reducing heat usage and by 1 % through reducing electricity usage from fiscal year 2002 to 2004. In particular, the effect is substantial in the Type 1 designated facilities, empirically demonstrating that the Energy Conservation Act is functioning well as a measure to mitigate climate change.

However, the amount of energy use in hotels in 2002 increased more than 38 % from 1990 levels. This result demonstrates that although the designated facilities were able to reduce their GHG emissions, the emissions of small and medium-sized facilities increased to a greater extent, which may have resulted in an overall increase in GHG emissions. Therefore, the amount of emission reductions by designated facilities may not be sufficient to meet the Kyoto Protocol reduction target.

This finding may present an important implication for domestic climate change mitigation policy. The Kyoto Protocol Target Achievement Plan required the designation of new facilities. For example, the Ministry of Economy, Trade and Industry encouraged small and medium-sized facilities to reduce GHG emissions by implementing a voluntary domestic CDM (Clean Development Mechanism)[19] for them. These small and medium-sized facilities may have substantial reduction potential.[20] To effectively reduce GHG emissions, it is necessary to include small and medium-sized facilities, which are not currently covered by the Energy Conservation Act.

In the future, we would like to consider the marginal abatement cost of reducing GHG emissions in the hotel industry. Because of data limitations, we were unable to estimate the marginal abatement cost. The analysis was conducted based on the short-term objective of compliance with the Kyoto Protocol. Therefore, it is impossible to determine whether it is cost effective to require hotels to improve their energy intensity at an annual rate of 1 %. For example, it may be cost effective

[19] Large firms support small and medium-sized firms in reducing CO_2 emissions. If small and medium-sized firms are able to reduce their CO_2 emissions successfully based on the support of large firms, then those reductions are considered large firms' reductions.

[20] However, according to some indications, domestic CDM was not initially implemented to a great extent. Hence, the transaction cost of including small and medium-sized firms in the scheme may be high and may thus be one of the future tasks to be addressed through policy making.

to require the hotel industry as a whole to improve its energy intensity by 1.5 % and to require a 0.5 % improvement in the steel industry. To extend the analysis, we must estimate the marginal abatement costs of various industries. Such efforts may clarify the optimal reduction level for the hotel industry. If differences in the marginal abatement cost across facilities are large, then it may be desirable to implement market-based instruments, such as emissions trading or a carbon tax, rather than imposing 1 % energy intensity improvement targets. Thus, marginal abatement costs provide crucial information for policy evaluations. This area should be addressed in future research.

Overall, we found that "comprehensive energy management" under the Energy Conservation Act was effective in Japan. Although this strategy represents an effective mitigation policy in Japan, can it be transferred to developing or emerging economies that are characterized by rapid growth in energy consumption? As we briefly noted in this chapter, the act does not contain particularly severe penalties for violations. Nevertheless, energy managers in the Japanese hotel industry successfully improved energy efficiency. One possibility is that social pressure or social norms are strong in Japanese society. Therefore, if a hotel is "publicly announced" as a violator, then the hotel may experience a decline in business. Another possibility is that the success of the act in Japan may result from the unique relationship between the government and the private sector. Strong ties with or trust in the government may give the private sector incentives to improve energy efficiency despite the lack of severe explicit penalties. These areas would also be interesting to investigate in future research.

Finally, the Energy Conservation Act was again partially amended, and the amendment became effective in 2009. Under this amendment, the transportation and food service industries were also covered. It would be useful to examine the effect of the amendment on those industries and whether the amendment has efficiently mitigated their GHG emissions.

Appendix 6.1 Quantile Regression

The aim of this chapter is to examine the effects of upgrading facility designations from Type 2 to Type 1 based on the amendment of the act in 2003. Using the results in Tables 6.4 and 6.5, we cannot robustly confirm the existence of positive effects of the upgrades on energy intensity. In particular, there may be substantial variations in the positive effects of the upgrade, as shown in Figs. 6.2 and 6.3. Therefore, we performed a quantile regression to capture the variation because the quantile regression methodology is useful when there are substantial variations across the left-hand side variables in an econometric analysis (Koenker and Hallock 2001).

Table 6.7 presents the results from quantile regressions for heat intensity and consumption. The results for heat intensity show that the coefficient of the Type 1 dummy is significantly negative at the 5 % level at the third quantile. This result indicates that upgrading from Type 1 from Type 2 improves heat energy intensity

Table 6.7 Quantile regression for heat

	Heat intensity						Heat consumption					
	Model A1						Model A2					
Type 1 dummy	−1.47		−2.64		−4.05	**	−116.7	**	−96.3	**	−118.9	
Heat 1	−0.87		−1.57		0.98		38.0		51.7		40.5	
Heat 2	3.57	*	2.63		−0.96		34.7		25.4		−21.5	
Heat 3	−0.78		1.33		2.84		18.9		68.1		75.2	
Heat 4	0.30		−0.55		−1.52		−89.5	*	−116.6	**	−101.6	
Heat 5	−2.81	**	−2.84		−3.78	*	−3.7		−18.8		−18.8	
Year dummy	4.50	***	4.92	*	4.25	***	130.1	***	101.5	***	124.8	*
Raw Sum	610		763		619		21,349		23,325		17,933	
Percentile	(0.25)		(0.50)		(0.75)		(0.25)		(0.50)		(0.75)	

Note: ***, ** and * indicate significance at the 1 %, 5 % and 10 % levels, respectively. The numbers of observations are 150 in Model A1 and 147 in Model A2. Although the climate and capacity variables are also used as explanatory variables in the estimation, they are omitted from this table

Table 6.8 Quantile regression for electricity

	Electricity intensity						Electricity consumption					
	Model A3						Model A4					
Type 1 dummy	−0.96		−1.82		−0.94		−204.7	***	−100.8		78.9	
Electricity 1	−0.79		−1.10		−0.52		−134.3		−48.5		12.0	
Electricity 2	1.32		1.89		0.90		140.0		93.1		2.5	
Electricity 3	−0.72		−0.22		−0.41		114.9		−23.3		−158.7	**
Electricity 4	0.54		0.30		0.37		−62.4		93.0		95.2	
Year dummy	2.37	***	1.68		2.22	***	217.0	***	149.5	**	194.1	***
Raw Sum	587		699		584		68,762		79,332		69,159	
Percentile	(0.25)		(0.50)		(0.75)		(0.25)		(0.50)		(0.75)	

Note: ***, ** and * indicate significance at the 1 %, 5 % and 10 % levels, respectively. The numbers of observations are 211 in Model A3 and 223 in Model A4. The climate and capacity variables are omitted from Table 6.7

by approximately 4 % at facilities with smaller improvements in heat energy intensity. Conversely, there are quantiles with small or insignificant of effects. Therefore, we observe substantial variation in the effect of the upgrade in the quantile regression results.

Next, we present the results for electricity by quantile regression. The results are presented in Table 6.8. Regarding the results of Model A3, the Type 1 dummy has no significant effect on electricity intensity. This result is consistent with the estimation results that are shown Table 6.1. The dummy is not found to have an effect in Model D in Table 6.5, whereas the coefficient of the dummy in Model A4 is significantly negative at the 1 % level at the first quantile. Therefore, the quantile regression reveals that there are variations in the effects across facilities.

References

Arimura TH, Iwata K (2007) The effects of "energy management" to fight global warming: comparative study with US energy star program. Stud Account 21:65–84 (In Japanese)

Arimura TH, Iwata K (2008) The CO_2 emission reduction under the Law Concerning the Rational Use of Energy: an empirical study of energy management in the Japanese hotel industry. Rev Environ Econ Policy Stud 1(1):79–89 (In Japanese)

Cabinet Public Relations Office (2005) Kyoto Protocol target achievement plan. (In Japanese) Available on http://www.kantei.go.jp/jp/singi/ondanka/kakugi/080328keikaku.pdf

Ito Y, Terao T (2005) Determinants of energy consumption in paper and pulp industry – analysis by using micro data of energy conservation law. In: Proceedings of society of environmental economics and policy studies 2005. Waseda University, Tokyo, pp 393–394 (In Japanese)

Koenker R, Hallock KF (2001) Quantile regression. J Econ Perspect 15(4):143–156

Ministry of the Environment Japan (2013) Greenhouse gas emission: measuring-reporting manual (Ver. 3.5) (in Japanese). Available on http://ghg-santeikohyo.env.go.jp/manual

Sugino M, Arimura TH (2011) The effects of voluntary action plans on energy-saving investment: an empirical study of the Japanese manufacturing sector. Environ Econ Policy Stud 13 (3):237–257

Sugiyama T, Kimura O, Noda F (2010) Energy conservation policy in Japan. Energy Forum. Tokyo (In Japanese)

Sugiyama T, Tanabe T (2002) Current status of energy conservation act – suggestion to prevent global warming. Socio-economic Research Center Discussion Paper. No. Y01934. (In Japanese)

The Energy Conservation Center Japan (2005) Baseline examination for promoting introduction of new energy 2004. (In Japanese) Available on http://www.eccj.or.jp/law06/com-judg/index.html

The Energy Data and Modeling Center (2005) Handbook of energy and economic statistics in Japan. The Energy Conservation Center Japan. Tokyo (In Japanese)

The Federation of Electric Power Companies of Japan (2006) Environmental action plan by the Japanese electric utility industry. (In Japanese) Available on https://www.fepc.or.jp/environment/warming/environment/pdf/2006.pdf

Wakabayashi M, Sugiyama T (2007) Japan's Keidanren voluntary action plan on the environment. In: Morgenstern RD, Pizer WA (eds) Reality check: the nature and performance of voluntary environmental programs in the United States, Europe, and Japan. Resources for the Future, Washington, DC, pp 43–63

Chapter 7
Economic Impacts of the GHG Tax on the Japanese Economy: Short-Term Analysis

Abstract The carbon tax is a market-based countermeasure for mitigating greenhouse gas (GHG) emissions that has received considerable attention around the world because a carbon tax is theoretically able to reduce GHG emissions at minimum cost. However, there are objections to the tax because the burdens of cost for specific energy-intensive industries, such as the coal and petroleum industries, may be large. This chapter conducts an economic impact analysis of a carbon tax proposal that was discussed by the Tokyo Tax Commission. The tax rate we examined was 3,415 Yen/t-CO_2. Specifically, we used an input-output framework to examine the economic impacts of this carbon tax proposal for each industry and the ways in which energy-intensive sectors may avoid or mitigate this cost burden.

In addition, we estimated the short-term impacts of the carbon tax proposal on those households that use the tax rate discussed by the Tokyo Tax Commission. An interesting feature of the proposed tax is the downstream taxation of electricity, whereby households directly pay the carbon tax for electricity usage. To examine the impact of this tax on households, we use the data from the "Family Income and Expenditure Survey" and link the price change determined from the Input-output model to the quantity listed in the survey.

We find that for some commodities, the price increase was greater than 5 %, and on average, the price increase was 1.2 %. For households, the results revealed that the tax burden was higher for low-income households and households living in cooler regions.

Keywords Carbon tax • Input-output analysis • Japanese Industry • Commodity price increase • Household expenditure • Regressive tax • Economic impact analysis

7.1 Introduction

Climate change policies and countermeasures have evolved around the globe. In Japan, the Japanese Business Federation's (Nippon Keidanren) voluntary action plan (VAP) for the environment has played a large role in the first commitment period of the Kyoto Protocol (Sugino and Arimura 2011). However, in the long run, there is no guarantee that this voluntary action plan will deliver large emission

T.H. Arimura, K. Iwata, *An Evaluation of Japanese Environmental Regulations*,
DOI 10.1007/978-94-017-9947-8_7

reductions. In addition, the VAP has been criticized for its potential inefficiency. Therefore, those policy instruments that use market mechanisms, such as emission trading and environmental taxes, have received attention. The 2009 manifesto of the Democratic Party of Japan listed three major climate change policy instruments: a domestic emission trading scheme (ETS), an environmental tax (a carbon dioxide tax as a global warming countermeasure) and feed-in tariffs (FIT). In December 2009, the "FY2010 Tax Reform proposal" was announced, which stated that a global warming countermeasure tax (carbon tax) at the national level would be introduced starting in FY2011 (Ministry of Finance Japan 2009).[1]

The process of implementing an environmental tax as a countermeasure for climate change has moved forward in recent years. Indeed, the national government has discussed the carbon tax as a source of revenue for the national government. Meanwhile, the Tokyo Tax Commission has discussed the implementation of a carbon tax as a local tax.

The Tokyo Tax Commission decided to conduct a simulation analysis to assess the economic impacts of the implementation of the carbon tax. Conducting a quantitative economic analysis at the discussion stage of environmental policy making was an epochal effort in the context of Japanese environmental policy. The assessment included (1) short-term economic impacts, (2) long-term economic impacts and (3) GHG reduction. The official analysis of the short-term economic impacts was conducted by Arimura Laboratory at Sophia University.

Computational general equilibrium models (CGE models) are often used to analyze the long-term economic effects and emission reductions of economic instruments such as carbon tax or emission trading schemes (Takeda et al. 2010). However, input-output models (IO models) are used to analyze the short-term burden of such policies (Sugino et al. 2013; Watanabe 2009; Shimoda and Watanabe 2006; Fujikawa 2002). This chapter will introduce the discussion held at the Tokyo Tax Commission and will provide an overview of the short-term impacts of the discussed tax rate.

The remainder of this chapter is organized as follows: the next section introduces the present Japanese energy-related taxation system along with the proposal discussed at the Tokyo Tax Commission, Sect. 7.3 introduces the methodology used in the economic analysis and Sect. 7.4 presents the results. In Sect. 7.5, the impacts of the environmental tax on the typical household will be presented. Section 7.6 concludes this chapter.

[1] An environmental tax was included in the "FS 2011 Tax Reform proposal," which was announced in December 2010 (Ministry of Finance Japan 2010).

7.2 The Energy-Related Taxation System in Japan and Key Points of the Proposed Carbon Tax

7.2.1 Energy-Related Taxation and the Discussion at the Tokyo Tax Commission

An environmental tax is an efficient policy method to internalize a negative externality. A "Carbon dioxide tax of global warming[2] as a countermeasure" (the local carbon hereafter) is an example of an environmental tax where the emission of carbon dioxide, which is the cause of climate change, is taxed. In practice, taxing fossil fuels is often considered when implementing a carbon tax because the emission of CO_2 for each fuel type is easily estimated using the carbon content of each fuel.

In Japan, fossil fuels have been subjected to various types of taxes, which we will explain based on Park (2009).[3] *The Petroleum and Coal Tax* imposes a charge rate of 2.04 Yen/l on gasoline diesel fuel, jet fuel, fuel oils A, B and C and kerosene. The Petroleum and Coal Tax also imposes a 1.08 Yen/l charge for liquefied petroleum gas and natural gas and 0.7 Yen/kg for coal. This tax was originally introduced to address energy security issues and was a source of revenue for the national government.

In addition to the Petroleum and Coal Tax, a *national gasoline tax* and a *local gasoline tax* are imposed on gasoline at the rates of 28.7 and 25.1 Yen/l, respectively. The national tax was originally used for the specified purpose of constructing roads for automobiles but later became part of the general budget. Similarly, diesel fuel is subject to the "light oil delivery tax," which is set at 32.1 Yen/l.[4] This tax was a source of revenue used by local governments to build roads.

Other taxes are also imposed on other fossil fuels. The *aviation fuel tax* was 26.0 Yen/l; this tax was collected to build airports. A *liquefied petroleum gas tax* was imposed on LPG at the rate of 17.5 Yen/kg, and half of the resulting revenue went to the national budget; the rest was added to the budgets of local governments. These funds were also used to build roads.

Finally, the *promotion of power resource development tax* was charged at the rate of 0.375 Yen/kwh to facilitate the construction of new power plants.

Because each tax was implemented to fulfill different objectives, the tax rates do not reflect the carbon content of the fuels. Therefore, the present energy tax system is inefficient at reducing CO_2 emissions. The Tokyo Tax Commission (2010) reports that the current energy tax system charges a per ton CO_2 tax rate of

[2] Whereas the Tokyo Tax Commission used "global warming" rather than "climate change," we use these words interchangeably in this chapter.

[3] See page 55 of Park (2009) for details.

[4] The "light oil delivery tax" is comprised of the principal tax rate of 15.0 Yen/l and a temporary tax rate of 17.1 Yen/l.

¥24,052 and ¥13,034 for gasoline and diesel fuel, respectively. In contrast, the per ton CO_2 tax rate for fuel oils A, B and C, coal and natural gas is ¥753, ¥291 and ¥400, respectively.

One should note the provisional tariff of the gasoline and diesel tax, as the Japanese gasoline tax and diesel tax consist of two parts: the original tax rate and a provisional tariff. These taxes originally went to special accounts to build roads and bridges. Some stakeholders argue that this provisional tariff part should be removed for consumers' benefit, but others are concerned that the removal of the provisional tariff may increase drivers' mileage and hence GHG emissions. Consequently, the DPJ proposed that the national government should keep the provisional tax rate and redefine it as a carbon tax.

However, the level of the proposed carbon tax rate at the national level was so low that it was not expected to meaningfully cut GHG emissions from the Japanese economy. In addition, the discussion of decentralizing various national government roles continued to move forward. The climate change policy was no exception. As a result, the Tokyo Tax Commission recognized that prefectural governments such as the Tokyo Metropolitan government must play an important role in countering climate change. The Tokyo Tax Commission discussed introducing a new carbon tax: the local tax (local carbon tax).

The Tokyo Tax Commission debated setting the level of the local carbon tax at 2,049 Yen/t-CO_2 (7,513 Yen/t-C). However, gasoline and diesel fuel for automobiles are exempted from this tax proposal.[5] The tax rate was set to make the tax revenue from the currently imposed energy-related taxes and the local carbon tax equal to 1.3 % of the GDP, which was the average tax on energy for OECD countries in 2004.

Although each prefectural government has the freedom to determine its own tax rate, we implicitly assume that other prefectures follow the Tokyo metropolitan government's lead. Therefore, our analysis assumes that other prefectural governments would introduce a tax with the same structure and at the same rate.

We calculate the tax rate for each type of energy based on the 2,049 Yen/t-CO_2 yields, and these rates are as follows: 5.552 Yen/l for fuel oils A, B and C, 5.101 yen/l for kerosene, 5.538 Yen/kg for natural gas and 4.769 Yen/kg for coal. This tax rate is also equivalent to 0.912 Yen/kwh for electricity. The details are shown in row 3 of Table 7.1.

A key factor in implementing a tax is simplifying the tax collection process. In other words, for fossil fuels, it is important to determine at which stage in the distribution process the tax should be imposed and collected. For each type of fuel, Fig. 7.1 depicts the stage where the taxes are collected for both the current energy-related tax and the proposal of the Tokyo Tax Commission. The Petroleum and Coal Tax is collected either at the border for imported fossil fuels or at the

[5] The midterm report by the Tokyo Tax Commission also reports the results using a tax rate of 3,415 Yen/t-CO_2 (12,522 Yen/t-C) as a reference case. The results for the reference case are presented in Sugino et al. (2012).

Table 7.1 The current energy-related tax and the discussed carbon tax as of March 2011

		Tax system as of 2011						Proposed carbon tax			
		The petroleum and coal tax (1)		Others (2)				Proposed tax rate (2,049 Yen/t-CO_2) (3)		Reference rate (3,415 Yen/t-CO_2) (4)	
Crude Oil		2.04	Yen/L	–				–		–	
Gasoline	Vehicle Use	2.04	Yen/L	(National/Local) Gasoline Tax	(Original Tax)	28.7	Yen/L	Redefinition of Provisional Tax as Carbon Tax			
					(Provisional Tariff)	25.1	Yen/L	25.100 Yen/L			
	Other Uses	2.04	Yen/L	–				4.757	Yen/L	7.929	Yen/L
Light Oil (Diesel Fuel)	Vehicle Use	2.04	Yen/L	Light Oil Delivery Tax	(Original Tax)	15.0	Yen/L	Redefinition of Provisional Tax as Carbon Tax			
					(Provisional Tariff)	17.1	Yen/L	17.100 Yen/L			
	Other Uses	2.04	Yen/L	–				5.297	Yen/L	8.828	Yen/L
Jet Fuel		2.04	Yen/L	Aviation Fuel Tax		26.0	Yen/L	5.046	Yen/L	8.410	Yen/L
Liquefied Petroleum Gas Tax	Vehicle Use	1.08	Yen/kg	Liquefied Petroleum Gas Tax		17.5	Yen/kg	6.145	Yen/kg	10.242	Yen/kg
	Other Uses	1.08	Yen/kg	–				6.145	Yen/kg	10.242	Yen/kg
Fuel Oil		2.04	Yen/L	–				5.552	Yen/L	9.254	Yen/L
Kerosene		2.04	Yen/L	–				5.101	Yen/L	8.502	Yen/L
Natural Gas		1.08	Yen/kg	–				5.538	Yen/kg	9.230	Yen/kg
Coal		0.7	Yen/kg	–				4.769	Yen/kg	7.949	Yen/kg
Electricity		–		Promotion of Power Resource Development Tax		0.375	Yen/kWh	0.912	Yen/kWh	1.520	Yen/kWh

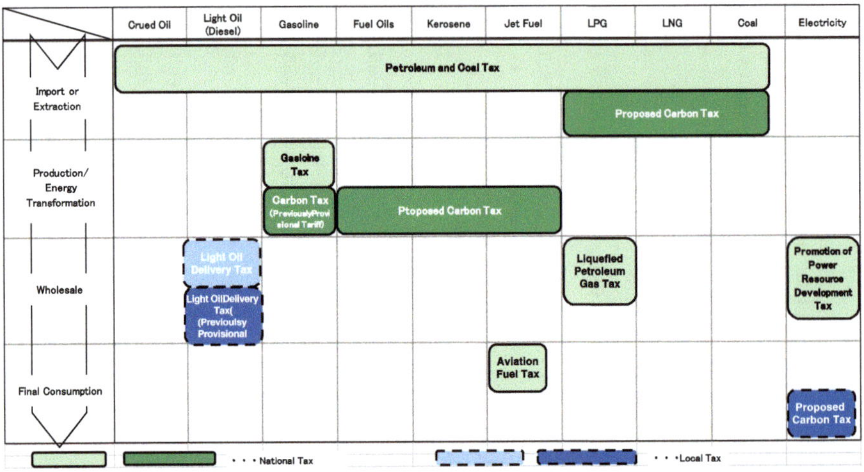

Fig. 7.1 Points of tax collection for the current taxation system and the global warming countermeasure tax as of March 2011

extraction site, and the gasoline tax is collected at the production or energy transformation stage (in each oil refinery). In addition, diesel oil, liquid petroleum gas (LPG) and electricity are taxed at the wholesale or retail stage. Finally, the aviation fuel tax is collected at the final consumption stage. As shown in Fig. 7.1, these taxes are collected at different stages of the distribution.

The discussion at the Tokyo Tax Commission had the following view on the collection of the new tax: the government should collect the tax for all fossil fuels other than diesel fuel at the border/extraction stage or at the production stage. As for diesel fuel, the temporary tariff of the Light Oil Delivery Tax should be collected at the wholesale/retail stage as a local tax. Furthermore, the taxation of electricity was considered as a local tax in the consumption stage. The Tokyo Tax Commission assumed that all prefectures will implement the same carbon tax as a local tax.

7.2.2 Key Issues in Carbon Taxation (the Global Warming Countermeasure Tax)

It is important to determine what economic issues are relevant when implementing the local carbon tax. First, we must ask whether the tax is able to internalize the negative externality caused by GHG emissions to optimize social welfare; if not, then it is impossible to set the rate of the carbon tax to improve social welfare. Measuring the marginal externality cost of GHG emissions is necessary to calculate the optimal tax rate to maximize social welfare. Whereas measuring this marginal externality cost is extremely difficult in climate change, implementing a carbon tax at any small tax rate will advance environmental policy because CO_2 emissions will be reduced due to the internalization of the externality.

Second, for a given level of emission reductions, the reduction in CO_2 will be efficiently achieved at minimum cost. Equalizing the per ton CO_2 tax rate among different fuel types will lead to economically efficient reductions (the equimarginal principle). As discussed above, the per ton CO_2 charge of the current tax system differs between fuels. In contrast, the proposal by the Tokyo Tax Commission tried to equalize the per ton CO_2 tax rate between fuels. The proposal may be considered an important advancement in environmental policy because the tax takes one step toward the equimarginal principle.

Third, we must ask to what extent the burden of the carbon tax may be imposed. The local carbon tax is considered to decarbonize firms and households in the long run and change the economy. However, in the short run, the new tax will create a burden because firms cannot easily change the technology they use (production processes), and households cannot easily change their consumption patterns. Therefore, the problem becomes determining how much burden firms and households are able to withstand.

Fourth, the problem of equality exists. The problem of equality becomes an issue if the new tax imposes a large burden on a specific group in the economy. For example, when considering an environmental tax on producers (firms), if the tax places a large burden on firms within a specific industry, these firms will raise the equality issue. One type of environmental tax is found to be regressive upon considering the effects of its implementation on households (Watanabe 2009). Therefore, if a new tax imposes a large burden on low-income households, then the tax is questionable.

If a new tax imposes a large burden on a specific group, gaining social consensus becomes difficult. Therefore, the institutional design of a tax must be well planned so that a specific industry or household group does not bear a disproportionally heavy burden. Sugino et al. (2013) analyze the effect of a tax exemption scheme for specific industries that may bear a large burden when a carbon tax is implemented. As shown in Sect. 7.4.1, the tax exemption scheme is not consistent with the principle of marginal cost equalization. From an economic perspective, this exemption scheme is undesirable because it reduces economic efficiency. However, reduced economic efficiency may be considered an indispensable cost to gain political acceptance of the carbon tax because equality cannot be ignored.

7.3 The Input-Output Model

7.3.1 Input-Output Analysis

In this section, we examine how a carbon tax affects our economy. First, a carbon tax will increase the prices of fossil fuels. Therefore, this tax will increase electricity prices because the generation of electricity uses fossil fuels in Japan. These increases in prices will raise the prices of various other goods, as fuel and electricity are required in their production. Some of these goods with increased prices serve as intermediate goods for final goods. In this way, a carbon tax is expected to increase the prices of almost all goods to certain degree. In investigating these complex effects on the prices, Input-output analysis (IO analysis) is a suitable method because it can incorporate the interactions between industries.

The IO analysis is conducted using data from the input-output table (IO table). In the IO table, the entire economy is divided into sectors, and economic transactions between industries are depicted. In the Japanese 2005 domestic IO table published by the Ministry of Internal Affairs and Communications, the basic classification divides the entire economy into 520 products and 407 industries. Other than the basic classification, there are three aggregated IO tables that have been published: a 190-sector classification table, a 108-sector classification table and a 34-sector classification table.

The IO table shows how many inputs are needed for each industry's production, such as intermediate goods, labor and capital, and where the outputs are allocated (i.e., intermediate demand and final demand). Table 7.2 depicts the IO table with two industries. For example, industry A uses 30 units of its own goods and 20 units of industry B's goods as intermediates. In addition, industry A uses 20 units of labor and 30 units of capital to produce a total of 100 units of goods. However, Industry A supplies Industries A and B with 50 units of intermediate goods in total, and the remaining 50 units are considered final goods.

Dividing the intermediate goods demanded by each industry by the total value of all of the shipments yields the "technical coefficient." The technical coefficient shows how many intermediate goods are needed to produce one unit of goods.

To assess the impacts of the implementation of a carbon tax, there are advantages of using IO analysis. First (as discussed above), the indirect effect of the tax caused by the cost increase of the intermediate goods may be analyzed using IO analysis. Second, the Japanese IO table includes the volumes of various fuels purchased by each industry, which is needed to analyze the detailed effects of taxation. Third, the effects of the tax, which include the cost increase, may be analyzed in detail for specific industries. Finally, Fig. 7.2 depicts the framework of the analysis that is discussed above.

Table 7.2 A simple input-output table with two industries

		Intermediate demand			
		Industry A	Industry B	Final demand	Total value of shipments
Intermediate Inputs	Industry A	30	20	50	100
	Industry B	20	40	60	120
Value Added	Labor	20	35		
	Capital	30	25		
Total Value of Shipments		100	120		

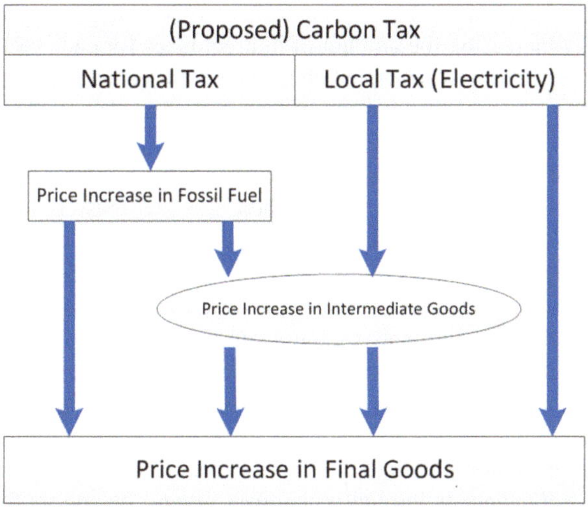

Fig. 7.2 An image of the analysis using the IO table

7.3.2 *Model*

We employ a price model that is often used in IO analysis to calculate the price increase of each product. This model allows us to calculate the price increase due to the implementation of a carbon tax. Furthermore, this model calculates the direct and indirect effects of the carbon tax.

For example, suppose that industry A raises the price of the produced good according to the cost burden of the carbon tax (the direct effect). This price increase raises the cost of other industries that use industry A's products as intermediate goods. Therefore, these industries will also raise the prices of their produced goods (the indirect effect). Equation 7.1 expresses the direct and indirect effects of the implementation of a carbon tax in vector form:

$$\mathbf{P} = \mathbf{V}'(\mathbf{I} - \mathbf{A})^{-1} \tag{7.1}$$

where $\mathbf{P}$ is the price vector, $\mathbf{V}$ is the total-value-added-coefficients vector, $\mathbf{I}$ is the identity matrix and $\mathbf{A}$ is the technical coefficients matrix. Finally, $(\mathbf{I} - \mathbf{A})^{-1}$ is the well-known Leontief inverse matrix, which captures the inter-industrial relationships.

The gross value added of each sector consists of the consumption expenditures outside households, the compensation of employees, the operation surplus, the depreciation of fixed capital, the indirect taxes and the current subsidies. The total value added coefficients are calculated by dividing the total value added of each sector by the total value of the shipments of each sector.

The proposed local carbon tax would change the total value added coefficients by increasing the amount of indirect taxes for each sector. These taxes on fuels are treated as new taxes that change the total value added coefficients. The exception is the tax on gasoline and diesel fuel. The Tokyo Tax Commission proposed to maintain the current tax rate for gasoline and diesel taxes.

For the sector classification, we use an aggregated table with 108 sectors[6] (i.e., 108 sector classifications). There are two reasons for using the aggregated table rather than 407 distinct classifications: first, the effects of the carbon tax on firms may become difficult to grasp if the disaggregated sector classification is used. For example, 407 classifications pin down the effects at the activity level, which may be so fine that the matrix cannot capture the effects at the firm level. In contrast, the aggregation of industries may underestimate the effects of the carbon tax on energy-intensive industries. Therefore, the 108-sector classification was chosen after considering the balance between the effect on the firms and properly evaluating the effects on industries.

IO analysis implicitly assumes several behaviors by producers of goods. First, the IO model shown above assumes that 100 % of the cost increase is passed to downstream industries in the form of higher prices. In addition, this model assumes that there are no substitution effects between goods. Therefore, the substitution of intermediate goods due to the carbon tax does not exist. This limitation also applies to fossil fuels, where coal, which has a high carbon content, is not substituted for other low carbon fuels, such as natural gas. Furthermore, because the analysis exists only over the short run, we assume that the diffusion of low-carbon technology is not stimulated by the carbon tax. Therefore, the results from the IO analysis may be considered to be the upper bound of the estimation. The simple IO model cannot calculate the CO_2 reduction brought by the carbon tax.

7.4 A Simulation Scenario and the Results of the IO Analysis

7.4.1 The Taxation Scenario

The Tokyo Tax Commission conducted a simulation analysis to observe the effects of the proposed local carbon tax. The tax rates considered by the commission were 2,049 Yen/t-CO_2 (7,513 Yen/t-C) and 3,415 Yen/t-CO_2 (12,522 Yen/t-CO_2).[7] We

[6] The Japanese IO table (basic classification) has 520 rows and 407 columns. Each row depicts products, whereas the columns depict industries. In other words, the IO table is rectangular, as there are more products than industries. The Leontief inverse matrix shown in Eq. 7.1 is calculated using the information in the 520 by 407 matrix. However, the number of rows and columns must be the same to obtain the Leontief inverse matrix.

[7] In both cases, the taxes on gasoline and diesel fuel for automobile use were exempted.

Table 7.3 Simulation scenarios

Scenario	Description
Scenario 1	The Implementation of the local carbon tax
Scenario 2 (Exemption Program)	Scenario 1 + tax exemption for energy-intensive sectors
Scenario 3	Scenario 1 + no tax on electricity

Table 7.4 Fuel types and industries that qualify for the tax exemption (Scenario 2)

Fuel	Industry
Coal/coke	Iron and Steel
Coal	Cement
Fuel oil A	Agriculture, Forestry and Fishery

will present the results for the 3,415 Yen/t-CO_2 tax rate, as the effects of a tax exemption program are clearer with a larger tax rate.[8]

If the carbon tax is set at the 3,425 Yen/t-CO_2 rate, then the tax rate for each fuel will become as follows: 9.254 Yen/l for fuel oils A, B and C, 8.502 Yen/l for kerosene, 9.230 Yen/kg for natural gas, 7.949 Yen/kg for coal and 1.520 Yen/kwh for electricity. Using these tax levels, we conducted a simulation analysis for the three scenarios in (Table 7.3). We assume that naphtha is exempted because tax systems often treat naphtha as untaxed.

In Scenario 1, all industries, including the energy-intensive industries, are subject to the environmental tax. However, energy-intensive industries are often considered to require special treatment because they have the possibility of losing international competitiveness (Sugino et al. 2013; Takeda et al. 2014). In Scenario 2, a tax exemption for energy-intensive industries is introduced along with the carbon tax. Table 7.4 lists the types of fuel and industry that qualify for the tax exemption. Similar exemption programs are discussed at the national level.

In Scenario 3, the carbon tax on electricity is exempted. Using the results from this scenario, the effects of the carbon tax on fossil fuels and electricity can be separated. In the simulation, we assume that fuel oils A, B and C, kerosene, natural gas and coal are taxed when they are consumed (combusted) at each industrial sector. This scenario is used to determine the difference in the price change resulting from the implementation of the tax exemption program.

[8] See Sugino et al. (2012) for comparable results using the 2,049 Yen/t-CO_2 tax rate.

7.4.2 Results

7.4.2.1 The Impacts of the Local Carbon Tax on the Entire Industry

Table 7.5 shows the price increase for the entire economy. Scenario 1 is the case in which there are no tax exemption programs, whereas Scenario 2 includes an exemption program. Without the tax exemption program, the average price increase is 1.183 %. However, when the tax exemption program is implemented, the average price increase is only 0.909 %. Therefore, the tax exemption program may reduce the price increase by approximately 0.28 %.

Table 7.6 shows the average price increase for the manufacturing sector and the transportation sector with and without the tax exemption program. The average price increase is 1.962 % for the entire manufacturing sector (Scenario 1). This price increase is higher than the price increase for the entire economy (Table 7.5), which suggests that the manufacturing sector bears more of the burden than other sectors. Nevertheless, the price increase falls to 1.376 % after the tax exemption program has been implemented for the entire manufacturing sector.

When the carbon tax is implemented, the price of railway transport increases by 0.949 %, the price of ocean transport increases by 5.079 %, the price of air transport increases by 3.621 % and the price of freight forwarding increases by 0.453 %. These results confirm that the carbon tax affects the transportation sector. The price increase for the respective transportation sectors are 0.919, 5.059, 3.594 and 0.443 % after the implementation of the tax exemption program. Therefore, the effects of the proposed tax exemption program are very limited for the transportation sector.

7.4.2.2 The Effects on Individual Industries

Table 7.7 shows the top 10 industries with the highest price increase due to the carbon tax without tax exemption programs.[9] Coal products (36.533 %), gas and heat (8.499 %), pig iron and crude steel (7.452 %), ocean transport (5.079 %) and steel products (4.677 %) are the top five industries with high price changes. Most of the top-ranked price-increasing industries use energy directly, i.e., the combustion of fossil fuels is employed in the production process.

Table 7.5 The average price increase after the implementation of the carbon tax

	Scenario 1 (without an exemption)	Scenario 2 (with an exemption)
Entire Economy	1.183 %	0.909 %

[9] See Appendix 7.1 for the price increase for each industry.

Table 7.6 Price increases for the manufacturing and transportation sectors

Sector		Scenario 1 (without an exemption)	Scenario 2 (with an exemption)
Manufacturing		1.962 %	1.376 %
Transportation	Railway Transport	0.949 %	0.919 %
	Ocean Transport	5.079 %	5.059 %
	Air Transport	3.621 %	3.594 %
	Freight Forwarding	0.453 %	0.443 %

Table 7.7 Price increases due to the carbon tax (the top 10 industries)

		Scenario 1 (without an exemption)		Scenario 2 (with an exemption)	
Sector ID		Price increase	Ranking	Price increase	Ranking
29	Coal Products	36.553 %	1	23.518 %	1
70	Gas and Heat Supply	8.499 %	2	8.460 %	2
37	Pig Iron and Crude Steel	7.452 %	3	2.542 %	8
81	Ocean Transport	5.079 %	4	5.059 %	3
38	Steel Products	4.677 %	5	2.031 %	13
82	Air Transport	3.621 %	6	3.594 %	4
39	Cast and Forged Steel Products	3.152 %	7	2.048 %	12
20	Chemical Fertilizers	3.132 %	8	3.029 %	5
21	Industrial Inorganic Chemicals	2.962 %	9	2.890 %	6
40	Other Iron and Steel Products	2.950 %	10	1.429 %	21

The results for 108 individual industries show that the energy-intensive industries' burden of the carbon tax is very large. Compared to the price increase for the entire economy (1.183 %), five industries experience a sharp price increase of more than 5 %. These results suggest that special treatment for the coal products industry may be needed in the short run due to the drastic price increase.

The results of the exemption program (Scenario 2) are also included in Table 7.7 to show the effects of the special treatment for energy-intensive industries. These results show that special treatment for the energy-intensive industries reduces the burden of the carbon tax to a certain degree. Indeed, the price increase for coal products falls from 36.553 to 23.518 %. Similarly, the price increase for pig iron and crude steel and steel products falls from 7.452 to 2.542 % and 4.677 to 2.031 %, respectively. However, the price increase for the gas and heat suppliers and ocean transport changes only slightly from 8.499 % and 5.079 to 8.460 % and 5.059 %, respectively.

The impact of the tax exemption program is limited for other highly ranked industries, such as air transport, chemical fertilizers and industrial inorganic chemicals. However, the impact of the tax exemption program is relatively large for industries that are directly affected, such as cast and forged steel products, cement and cement products, fisheries and agriculture.

These results suggest that special treatment for the energy-intensive industries reduces their burden from the carbon tax. However, for some industries, the effects were limited. As for coal products, the price increase only decreases to 23.518 %, which implies that further "special treatment" is needed in the short run.

7.4.3 *The Effects of National and Local Taxes*

The discussion at the Tokyo Tax Commission assumed that the tax on electricity was collected as a local tax. We thus decompose the effect of the carbon tax into two parts: the effect of the tax on electricity usage, and the effect of the tax on fossil fuels (case 3).

Table 7.8 shows the effects of the tax on other energy sources and the effects of the tax on electricity for the top 10 industries with a high price increase due to the carbon tax. The coal products industry has the highest price increase of 36.552 % due to the carbon tax. However, the effect of this tax on electricity is an increase of only 0.573 %. The results for other industries with high price increases, such as the gas and heat supply industry and the pig iron and crude steel industry, show similar results.

Table 7.8 Decomposition of the price increases (top 10 industries)

		Impacts of the carbon tax (total)		Impacts of the tax on fossil fuel		Impacts of the electricity tax	
Sector ID	Industry	Price increase	Rank	Price increase	Rank	Price increase	Rank
29	Coal Products	36.553 %	1	35.980 %	1	0.573 %	23
70	Gas and Heat Supply	8.499 %	2	8.050 %	2	0.449 %	37
37	Pig Iron and Crude Steel	7.452 %	3	6.321 %	3	1.131 %	3
81	Ocean Transport	5.079 %	4	4.920 %	4	0.159 %	95
38	Steel Products	4.677 %	5	3.812 %	5	0.865 %	6
82	Air Transport	3.621 %	6	3.359 %	6	0.262 %	73
39	Cast and Forged Steel Products	3.152 %	7	2.034 %	11	1.118 %	4
20	Chemical Fertilizers	3.132 %	8	2.438 %	7	0.694 %	11
21	Industrial Inorganic Chemicals	2.962 %	9	1.270 %	18	1.692 %	1
40	Other Iron and Steel Products	2.950 %	10	2.306 %	8	0.644 %	14

Table 7.9 The effects of electricity taxes on prices (top 10 industries)

Sector ID	Sectors	Price increase	Rank
21	Industrial Inorganic Chemicals	1.692 %	1
6	Metallic Ores	1.137 %	2
37	Pig Iron and Crude Steel	1.131 %	3
39	Cast and Forged Steel Products	1.118 %	4
41	Non-ferrous Metals	1.010 %	5
38	Steel Products	0.865 %	6
25	Synthetic Fibers	0.861 %	7
84	Storage Facility Service	0.789 %	8
42	Non-ferrous Metal Products	0.753 %	9
78	Railway Transport	0.745 %	10

In case 3, the average price increase is 0.812 %. However, the average price increase is 1.183 % if the carbon tax is imposed on all energy sources, including electricity. Therefore, the tax on energy sources excluding electricity contributes approximately 68.6 % (=0.812/1.183) of the total price increase if all of the energy sources are taxed. In contrast, the taxation of electricity leads to an average price increase of 0.370 %, or 31 % of the total price increase caused by the carbon tax.

Table 7.9 presents the top 10 industries with high price increases caused by the taxation of electricity. The top five industries that experience a high price increase are industrial inorganic chemicals (1.692 %), metallic ores (1.137 %), pig iron and crude steel (1.131 %), cast and forged steel products (1.118 %) and non-ferrous metals (1.010 %).

Among other things, the above results have two implications. First, the effect of electricity taxation on the average price is small compared to the effects of taxes on fossil fuels. Second, the effects of electricity taxation between industries are not as diverse as the effects of taxes on fossil fuels. In other words, the tax on electricity does not put the entire burden on specific industries.

7.4.4 *The Effects of Carbon Taxation by Fuel Type*

Next, the price increase by fuel type was estimated (Table 7.10). In the simulations, we separately simulated a tax on gasoline, other petroleum products (kerosene, heavy oil, aviation fuel and petroleum gas), natural gas and coal. In doing so, we examined two cases: with and without the tax exemption program. The average price increase for the entire manufacturing industry without the tax exemption program was 1.962 %.

The tax implemented on coal and electricity usage accounted for 0.970 and 0.550 % of the price increase, respectively. The price increase caused by the two energy sources adds up to 1.520 %, or 77 % of the total price increase. Therefore, these two energy sources account for a large portion of the increase in the average

Table 7.10 Price increases by fuel type

		Gasoline	Petroleum products	Gas	Coal	Electricity	Total
All Industries	Without an Exemption (Scenario 1)	0.004 %	0.267 %	0.084 %	0.457 %	0.370 %	1.183 %
Manufacturing		0.004 %	0.340 %	0.099 %	0.970 %	0.550 %	1.962 %
All Industries	With an Exemption (Scenario 2)	0.004 %	0.255 %	0.084 %	0.197 %	0.370 %	0.910 %
Manufacturing		0.004 %	0.332 %	0.099 %	0.394 %	0.550 %	1.378 %

price. However, if the tax exemption program is implemented, the average increase in price for the manufacturing sector falls from 1.962 to 1.378 %.

Inasmuch as the exemption program targets coal and other petroleum products, the coal tax falls from 0.970 to 0.394 %. In other words, the tax exemption program is effective in constraining the price increase.

7.5 The Effects on Households

7.5.1 A Framework of the Impacts on the Household

A carbon tax has impacts on the household as well. In this section, we analyze the impact of the local carbon tax on the typical household using the results obtained from the input-output analysis. We use the method employed by Shimoda and Watanabe (2006) and Fujikawa (2002) and estimate the impact of the price increase caused by the carbon tax on households. Rather than estimate the "average household," we estimate the impacts on different households according to their income levels.

In the analysis, we use the expenditure data from the "Family Income and Expenditure Survey (FIES)" along with the price increase calculated from the IO analysis. Table 7.14 in the appendix shows the correspondence between the items in the FIES and the industry classification of the IO table. Furthermore, we restrict the data to households with two or more family members. Consequently, we are able to estimate the effects of the carbon tax on households with different income levels.

In addition to the price increase of non-energy goods and services, we estimate the effects of the price increase of energy (i.e., electricity), kerosene and city gas.

The additional payments for electricity use and kerosene caused by the carbon tax are calculated by multiplying the per unit carbon tax by the average fuel/electricity consumption per household. The average price of electricity is ¥22.25 per kwh, whereas kerosene costs ¥64.02 per liter in the FIES.

In contrast to the cases of electricity and kerosene, the FIES does not provide detailed city gas prices. Consequently, we estimate the average price of city gas by using the figures presented by the "Retail Price Survey" under the category of "Gas Expenditure." We calculate the average city gas price by taking the average of the city gas prices in the 47 prefectural capitals because the price of city gas differs between regions. The calculated city gas price was ¥163.08 per cubic meter.[10]

[10]The Retail Price Survey reports prices of goods every year in three different categories: prefectural capital and cities with more than 150,000 people, cities with between 50,000 and 150,000 people and cities with fewer than 50,000 people. We used the prices listed in the table for prefectural capitals and cities with more than 150,000 people to calculate the average city gas price.

7.5.2 *The Impacts of the Carbon Tax on Household Expenditures with Different Levels of Income*

This subsection analyzes the effects of the carbon tax on households by income level. Households are classified into 18 categories according to their annual income by the FIES.

The column titled "Product increase (D)" in Table 7.11 shows the indirect effects of the carbon tax without (scenario 1) and with (scenario 2) the tax exemption program. In case 1, the expenditures are estimated to increase by ¥2,522 per month for the average household. In case 2, these expenditures are estimated to increase by ¥2,254 per month.

Due to the price increases and changes in the relative prices of goods, in reality households may reduce their consumption or change their consumption bundle. However, our analysis assumes that consumers neither change their consumption bundle nor reduce their consumption in response to the higher prices. Therefore, the estimated figures may be considered upper-bound values of the increases due to the carbon tax.

Column E in Table 7.11 shows the increase in the utility costs caused by the carbon tax. If households do not respond to the increase in utility costs by consuming less, then the increase in the utility cost for the average household is estimated to be ¥1,163 per month. In reality, households are expected to respond to higher utility prices by consuming less electricity and fuel. Therefore, the actual increase in household expenditures will be less than ¥1,163 per month.

Table 7.11 also shows that the increased prices for products and services have a greater effect on households than the increased utility price. This result implies that it is important to analyze the impact of higher product and service prices along with the direct tax payment that is included in utilities.

The total impact of the carbon tax on the typical household is shown in column C, which is the sum of column D and column E. The increase in the household expenditures without the exemption program for the average household is estimated to be ¥3,685 per month. In contrast, this household's expenditure increase will be reduced to ¥3,416 per month with the tax exemption program. The two estimates above assume that imported goods will also be taxed equivalently with their domestic counterparts. However, if we assume that the imported goods will be exempted from the carbon tax, then the increase in household expenditures for the average household without and with the tax exemption program will be ¥3,077 and ¥2,922 per month, respectively.

We can estimate the size of the burden of the carbon tax borne by the household. The ratio of the increase in household expenditures due to the carbon tax to the household's original expenditures is shown in column F of Table 7.11. The average household will experience a 1.26 % increase in its total household expenditures without the exemption program. If the carbon tax includes an exemption program, then the ratio falls to 1.17 %.

Table 7.11 The effects of the carbon tax on household expenditures by income

Annual household income (10 thousand yen)	Average monthly household expenditure (Yen) (A)	Annual household income (10 thousand Yen) (B)	Monthly increase in household expenditure (Yen)					Increase rate in household expenditure (%) (F)		
			Total increase (C) = (D) + (E)		Expenditure increase (products/services) (D)		Utility cost increase (E) (electricity, gas, kerosene)	Total Increase (%) (E)/(F)		Utility cost only (%) (E)/(A)
			Without Exemption	With Exemption	Without Exemption	With Exemption		Without Exemption	With Exemption	
Average	291,737	629	3,685	3,416	2,522	2,254	1,163	1.263 %	1.171 %	0.399 %
Less than 200	119,016	157	2,127	1,981	1,237	1,091	890	1.787 %	1.664 %	0.748 %
200–250	186,863	226	2,777	2,576	1,748	1,547	1,029	1.486 %	1.378 %	0.551 %
250–300	188,470	275	2,855	2,643	1,790	1,578	1,065	1.515 %	1.402 %	0.565 %
300–350	216,890	323	3,073	2,844	2,008	1,779	1,065	1.417 %	1.311 %	0.491 %
350–400	229,492	373	3,161	2,925	2,089	1,854	1,072	1.377 %	1.275 %	0.467 %
400–450	249,715	423	3,312	3,073	2,219	1,980	1,093	1.326 %	1.231 %	0.438 %
450–500	259,040	473	3,417	3,169	2,289	2,042	1,127	1.319 %	1.223 %	0.435 %
500–550	273,659	522	3,517	3,267	2,378	2,128	1,139	1.285 %	1.194 %	0.416 %
550–600	283,325	573	3,581	3,324	2,449	2,191	1,133	1.264 %	1.173 %	0.400 %
600–650	303,952	621	3,726	3,455	2,592	2,322	1,134	1.226 %	1.137 %	0.373 %
650–700	315,286	672	3,845	3,565	2,671	2,392	1,174	1.219 %	1.131 %	0.372 %
700–750	325,761	722	4,022	3,738	2,810	2,526	1,212	1.235 %	1.147 %	0.372 %
750–800	330,887	772	4,026	3,742	2,808	2,524	1,219	1.217 %	1.131 %	0.368 %
800–900	360,895	844	4,223	3,917	2,997	2,691	1,226	1.170 %	1.085 %	0.340 %
900–1,000	404,250	944	4,585	4,247	3,270	2,932	1,315	1.134 %	1.050 %	0.325 %
1,000–1,250	401,728	1,098	4,703	4,361	3,337	2,995	1,366	1.171 %	1.086 %	0.340 %
1,250–1,500	432,944	1,354	4,869	4,507	3,528	3,166	1,341	1.125 %	1.041 %	0.310 %
More than 1,500	529,475	1,968	5,824	5,392	4,290	3,858	1,534	1.100 %	1.018 %	0.290 %

Column F of Table 7.11 also shows that low-income households have higher ratios than high-income households. For example, the ratio for households with annual incomes between ¥2 million and ¥2.5 million is 1.49 %, whereas the ratio is 1.12 % for households with annual income between ¥12.5 million and ¥15 million. These figures suggest that the carbon tax is regressive, if only slightly.

7.5.3 *The Impacts of the Carbon Tax on Household Expenditures in Different Regions*

Next, we estimated the effects of the carbon tax on household dwellings in different regions of Japan. The results are shown in Table 7.12. Column D of Table 7.12 shows the increase of household expenditures that are caused by higher prices for goods and services. Column E shows the additional household expenditures that are caused by higher utility costs, whereas column C adds the figures in columns D and E for each region.

The results indicate that the average household living in Hokkaido, the northernmost island in Japan, will bear the greatest burden, facing a 1.45 % increase in total household expenditures. Although households in Hokkaido have lower additional expenditures for goods and services compared with other regions, they have higher utility costs of ¥1,618 per month.

In contrast, for households in Kyushu, the southernmost island in Japan, the utility cost increase is only ¥979 per month, which represents approximately 60 % of the increase of Hokkaido. The difference in additional household expenditures between Hokkaido and other regions may be explained by the differences in utility costs, which may be attributed to differences in heating requirements during the winter.

In addition, column E of Table 7.12 shows the contribution of electricity and kerosene to these additional utility costs. Column E implies that the primary factor of the unequal burden of utility costs arises from kerosene used for heating purposes relative to other purposes. The increase in kerosene costs is higher for Hokkaido and the Tohoku region, which is located in the northern part of Honshu, the main island of Japan. In addition, high electricity costs are expected in the Hokuriku, Shikoku and Chugoku regions.

As for the Tokyo metropolitan area, the average household will pay 1.17 % more, which is lower than the national average of 1.26 %. This difference may be attributed to the relatively mild weather in the region. In addition, the dependence of the public on automobiles is the weakest in the Tokyo metropolitan area due to the development of public transportation.

Table 7.12 The effects of the carbon tax on household expenditures by region

Region	Average monthly household expenditure (yen), (A)	Monthly increase in household expenditure (yen)							Increase in household expenditure (%)	
		Total (C) = (D) + (E)		Product increase (D)		Utility costs (E)				
		Without exemption	With exemption	Without exemption	With exemption	Cost increase	(Electricity)	(Kerosene)	Without exemption	With exemption
National Level	291,737	3,685	3,417	2,522	2,254	1,163	659	181	1.26 %	1.17 %
Hokkaido	260,262	3,772	3,530	2,154	1,912	1,618	609	795	1.45 %	1.36 %
Tohoku	278,708	3,946	3,681	2,423	2,158	1,523	683	534	1.42 %	1.32 %
Kanto	300,408	3,692	3,420	2,606	2,334	1,086	630	109	1.23 %	1.14 %
Hokuriku	318,997	4,097	3,793	2,690	2,386	1,407	790	323	1.28 %	1.19 %
Tokai	300,223	3,828	3,558	2,636	2,366	1,192	703	133	1.28 %	1.19 %
Kinki	295,288	3,677	3,398	2,590	2,311	1,087	667	78	1.25 %	1.15 %
Cyugoku	293,963	3,686	3,414	2,485	2,213	1,201	759	162	1.25 %	1.16 %
Shikoku	294,759	3,604	3,329	2,469	2,194	1,135	740	138	1.22 %	1.13 %
Kyushu	263,046	3,245	3,007	2,266	2,028	979	568	116	1.23 %	1.14 %
Okinawa	233,032	2,843	2,650	1,862	1,669	981	653	81	1.22 %	1.14 %
Kanto excluding Tokyo	308,567	3,750	3,470	2,685	2,405	1,065	621	80	1.22 %	1.12 %
Cyukyo	292,971	3,758	3,493	2,606	2,341	1,152	694	93	1.28 %	1.19 %
Keihanshin	298,286	3,714	3,432	2,640	2,358	1,074	651	58	1.25 %	1.15 %
Kitakyusyu/ Fukuoka	280,198	3,461	3,205	2,414	2,158	1,047	627	114	1.24 %	1.14 %
Tokyo Met-ropolitan Area	326,263	3,823	3,536	2,784	2,497	1,039	635	30	1.17 %	1.08 %

Note: The four rows at the bottom shows the results at the aggregate regional levels

7.6 Conclusion

This chapter presented a brief summary of the short-term economic impact analysis of the carbon tax discussed at the Tokyo Tax Commission. An input-output model was used to estimate the effects of the carbon tax on the economy as well as the impacts on individual industries. In addition, the IO model was used to estimate the effects of the exemption program for industries that will be affected the most, i.e., carbon-intensive industries. The results on the tax exemption program will be useful in designing the specific structure of a carbon tax.

This chapter also examined the impact of the carbon tax on household expenditures by using the results obtained from the IO analysis. These results indicate that the impacts of higher good prices are greater than the impacts of higher utility prices. In addition, the importance of the IO analysis was shown in the household expenditure analysis. The results also showed that the carbon tax is regressive, although the magnitude is limited.

In general, tax exemption programs for energy-intensive industries reduce the economic efficiency of the carbon tax. Therefore, any tax exemption should be tentatively implemented only in the early stages of the carbon tax. In reality, any implementation of the carbon tax will be difficult without such a tax exemption program. Therefore, this inefficiency of the exemption program may be considered an inevitable short-run cost. The inefficiency's size may be estimated using computational general equilibrium models (CGE models), which may address the long-run costs as well. In other words, by comparing the results from different scenarios of tax exemption programs, the relative size of the inefficiency may be estimated. For example, Fischer and Fox (2007) and Takeda et al. (2014) estimated this type of inefficiency from a special treatment of the energy-insensitive sectors of the Japanese economy under potential cap and trade schemes. Quantitative analysis using economic models may provide valuable information that policy makers require to balance the economic efficiency of the carbon tax with the burden it places on households.

Appendix 7.1

Table 7.13 Price increases due to the carbon tax by sector

Industry(108 sector classification)			Scenario 1 (without an exemption)		Scenario 2 (with an exemption)	
			Price increase	Ranking	Price increase	Ranking
Agriculture, forestry and fishery/Mining	1	Crop cultivation	0.760 %	66	0.472 %	84
	2	Livestock	0.651 %	76	0.570 %	78
	3	Agricultural services	0.613 %	81	0.594 %	75
	4	Forestry	0.374 %	92	0.312 %	99
	5	Fisheries	2.324 %	14	0.800 %	52
	6	Metallic ores	1.597 %	23	1.552 %	20
	7	Non-metallic ores	0.939 %	49	0.907 %	46
	8	Coal mining, crude petroleum and natural gas	0.874 %	56	0.823 %	50
Manufacturing	9	Foods	0.883 %	53	0.738 %	61
	10	Beverage	0.649 %	77	0.597 %	73
	11	Feeds and organic fertilizer, n.e.c.	0.828 %	61	0.715 %	64
	12	Tobacco	0.163 %	104	0.144 %	106
	13	Textile products	1.768 %	21	1.731 %	17
	14	Wearing apparel and other textile products	0.974 %	47	0.949 %	40
	15	Timber and wooden products	0.628 %	79	0.596 %	74
	16	Furniture and fixtures	0.876 %	54	0.765 %	59
	17	Pulp, paper, paper-board, building paper	2.316 %	15	2.278 %	10
	18	Paper products	1.095 %	39	1.073 %	30
	19	Printing, plate making and book binding	0.660 %	73	0.645 %	69
	20	Chemical fertilizer	3.132 %	8	3.029 %	5
	21	Industrial inorganic chemicals	2.962 %	9	2.890 %	6
	22	Petrochemical basic products	1.696 %	22	1.656 %	19
	23	Organic chemical products (except Petrochemical basic products)	2.568 %	13	2.391 %	9
	24	Synthetic resins	1.800 %	20	1.719 %	18
	25	Synthetic fibers	2.723 %	12	2.647 %	7
	26	Medicaments	0.694 %	70	0.666 %	67
	27	Final chemical products, n.e.c.	1.221 %	32	1.170 %	26
	28	Petroleum refinery products	0.864 %	57	0.829 %	48

(continued)

Table 7.13 (continued)

Industry(108 sector classification)			Scenario 1 (without an exemption)		Scenario 2 (with an exemption)	
			Price increase	Ranking	Price increase	Ranking
	29	Coal products	36.553 %	1	23.518 %	1
	30	Plastic products	1.162 %	34	1.117 %	27
	31	Rubber products	1.121 %	36	1.047 %	33
	32	Leather, fur skins and miscellaneous leather products	0.616 %	80	0.586 %	77
	33	Glass and glass products	1.835 %	19	1.808 %	15
	34	Cement and cement products	2.735 %	11	0.935 %	42
	35	Pottery, china and earthenware	2.196 %	16	2.164 %	11
	36	Other ceramic, stone and clay products	2.041 %	17	1.933 %	14
	37	Pig iron and crude steel	7.452 %	3	2.542 %	8
	38	Steel products	4.677 %	5	2.031 %	13
	39	Cast and forged steel products	3.152 %	7	2.048 %	12
	40	Other iron or steel products	2.950 %	10	1.429 %	21
	41	Non-ferrous metals	1.853 %	18	1.796 %	16
	42	Non-ferrous metal products	1.412 %	29	1.368 %	23
	43	Metal products for construction and architecture	1.510 %	25	1.106 %	28
	44	Other metal products	1.416 %	28	1.070 %	31
	45	General industrial machinery	0.989 %	46	0.798 %	53
	46	Special industrial machinery	0.892 %	52	0.726 %	63
	47	Other general machines	1.118 %	37	0.884 %	47
	48	Machinery for office and service industry	0.852 %	58	0.764 %	60
	49	Electrical devices and parts	0.927 %	51	0.794 %	54
	50	Applied electronic equipment and electric measuring instruments	0.683 %	72	0.644 %	70
	51	Other electrical equipment	1.049 %	44	0.977 %	38
	52	Household electric appliances	0.875 %	55	0.771 %	56
	53	Household electronics equipment	0.816 %	62	0.765 %	58

(continued)

Table 7.13 (continued)

Industry(108 sector classification)			Scenario 1 (without an exemption)		Scenario 2 (with an exemption)	
			Price increase	Ranking	Price increase	Ranking
	54	Electronic computing equipment and accessory equipment of electronic computing equipment	0.766 %	64	0.728 %	62
	55	Semiconductor devices and Integrated circuits	1.098 %	38	1.059 %	32
	56	Other electronic components	1.053 %	42	1.007 %	35
	57	Passenger motor cars	1.060 %	41	0.930 %	43
	58	Other cars	1.066 %	40	0.937 %	41
	59	Motor vehicle parts and accessories	1.186 %	33	1.017 %	34
	60	Ships and repair of ships	1.568 %	24	1.202 %	24
	61	Other transportation equipment and repair of transportation equipment	1.140 %	35	0.990 %	36
	62	Precision instruments	0.763 %	65	0.713 %	65
	63	Miscellaneous manufacturing products	0.691 %	71	0.630 %	71
Services and Others	64	Reuse and recycling	1.418 %	27	1.403 %	22
	65	Building construction	0.655 %	74	0.507 %	83
	66	Repair of construction	0.729 %	67	0.569 %	79
	67	Public construction	1.425 %	26	0.984 %	37
	68	Other civil engineering and construction	1.052 %	43	0.773 %	55
	69	Electricity	1.320 %	30	1.095 %	29
	70	Gas and heat supply	8.499 %	2	8.460 %	2
	71	Water supply	1.223 %	31	1.192 %	25
	72	Waste management service	0.842 %	60	0.823 %	49
	73	Commerce	0.348 %	96	0.338 %	95
	74	Finance and insurance	0.160 %	105	0.155 %	104
	75	Real estate agencies and rental services	0.302 %	100	0.293 %	100
	76	House rent	0.116 %	107	0.104 %	107
	77	House rent (imputed house rent)	0.047 %	108	0.039 %	108
	78	Railway transport	0.949 %	48	0.919 %	44
	79	Road transport (except transport by private cars)	0.388 %	90	0.377 %	90

(continued)

Table 7.13 (continued)

Industry(108 sector classification)			Scenario 1 (without an exemption)		Scenario 2 (with an exemption)	
			Price increase	Ranking	Price increase	Ranking
	80	Self-transport by private cars	0.631 %	78	0.598 %	72
	81	Water transport	5.079 %	4	5.059 %	3
	82	Air transport	3.621 %	6	3.594 %	4
	83	Freight forwarding	0.453 %	87	0.443 %	87
	84	Storage facility service	0.936 %	50	0.913 %	45
	85	Services relating to transport	0.487 %	85	0.468 %	85
	86	Communication	0.250 %	101	0.242 %	101
	87	Broadcasting	0.369 %	94	0.358 %	93
	88	Information services	0.224 %	102	0.217 %	102
	89	Internet based services	0.328 %	98	0.317 %	97
	90	Image information, character information production and distribution	0.520 %	84	0.508 %	82
	91	Public administration	0.377 %	91	0.362 %	92
	92	Education	0.321 %	99	0.313 %	98
	93	Research	0.787 %	63	0.769 %	57
	94	Medical service and health	0.463 %	86	0.449 %	86
	95	Social security	0.397 %	89	0.378 %	89
	96	Nursing care	0.362 %	95	0.340 %	94
	97	Other public services	0.341 %	97	0.329 %	96
	98	Advertising services	0.373 %	93	0.364 %	91
	99	Goods rental and leasing services	0.178 %	103	0.167 %	103
	100	Repair of motor vehicles and machine	0.654 %	75	0.588 %	76
	101	Other business services	0.157 %	106	0.152 %	105
	102	Amusement and recreational services	0.563 %	83	0.549 %	81
	103	Eating and drinking places	0.711 %	69	0.646 %	68
	104	Accommodations	0.845 %	59	0.800 %	51
	105	Cleaning, barber shops, beauty shops and public baths	0.568 %	82	0.558 %	80
	106	Other personal services	0.403 %	88	0.389 %	88
	107	Office supplies	0.999 %	45	0.968 %	39
	108	Activities not elsewhere classified	0.724 %	68	0.687 %	66
	–	The average of overall industry	1.183 %	–	0.909 %	–

Table 7.14 Sector correspondences between *Family Income and Expenditure Survey and Input-Output Table*

Sector (family income and expenditure survey)			Sector (input-output table)
Food	Cereals	Rice	Crop cultivation
		Bread	Crop cultivation
		Noodles	Crop cultivation
		Other cereals	Crop cultivation
	Fish & shellfish	Raw fish & shellfish	Fisheries
		Raw fish & shellfish	Foods
		Fish-paste products	Foods
		Other processed fish	Foods
	Meat	Raw meat	Livestock
		Processed meat	Foods
	Dairy products & eggs		Foods
	Vegetables & seaweeds		Crop cultivation
	Fruits		Crop cultivation
	Oils, fats & seasonings		Foods
	Cakes & candies		Foods
	Cooked food		Foods
	Beverages		Beverage
	Alcoholic beverages		Beverage
	Meals outside the home		Eating and drinking places
Housing	Rents for dwelling & land		House rent
	Repairs & maintenance		Building construction
Fuel, light & water charges	Electricity		Electricity
	Gas		Gas and heat supply
	Other fuel & light		Petroleum refinery products
	Water & sewerage charges		Water supply
Furniture & household utensils	Household durable goods	Durable goods assisting housework	Household electric appliances
		Heating & cooling appliances	Household electric appliances
		General furniture	Furniture and fixtures
	Interior furnishings & decorations		Furniture and fixtures
	Bedding		Furniture and fixtures
	Domestic utensils		Miscellaneous manufacturing products
	Domestic non-durable goods		Miscellaneous manufacturing products
	Domestic services		Other personal services
Clothing & footwear	Japanese clothing		Wearing apparel and other textile products
	Clothing		Wearing apparel and other textile products
	Shirts & sweaters		Wearing apparel and other textile products
	Underwear		Wearing apparel and other textile products
	Cloth & thread		Textile products
	Other clothing		Wearing apparel and other textile products

(continued)

Table 7.14 (continued)

Sector (family income and expenditure survey)			Sector (input-output table)
	Footwear		Leather, fur skins and miscellaneous leather products
	Services related to clothing		Other personal services
Medical care	Medicines		Medicaments
	Health fortification		Medicaments
	Medical supplies & appliances		Medical service and health
	Medical services		Medical service and health
Transportation & communication	Public transportation		Railway transport
	Private transportation	Purchase of vehicles	Passenger motor cars
		Purchase of bicycles	Miscellaneous manufacturing products
		Maintenance of vehicles	Petroleum refinery products
	Communication		Communication
Education			Education
Culture & recreation	Recreational durable goods		Household electronics equipment
	Recreational goods		Miscellaneous manufacturing products
	Books & other reading materials		Printing, plate making and book binding
	Recreational services	Accommodation services	Accommodations
		Package tours	Accommodations
		Lesson fees	Education
		Other recreational services	Amusement and recreational services
Other consumption expenditures	Miscellaneous	Personal care services	Other personal services
		Personal care goods	Final chemical products, n.e.c.
		Personal effects	Wearing apparel and other textile products
		Tobacco	Tobacco
		Other miscellaneous	Miscellaneous manufacturing products
	Pocket money (of which, detailed uses unknown)		Eating and drinking places
	Social expenses	Food	Eating and drinking places
		Furniture & household utensils	Furniture and fixtures
		Clothing & footwear	Textile products
		Culture & recreation	Amusement and recreational services
		Other goods & services	Other personal services
		Money gifts	Other personal services
		Other social expenses	Eating and drinking places
	Remittance		House rent

References

Fischer C, Fox AK (2007) Output-based allocation of emission permits for mitigating tax and trade interactions. Land Econ 83(4):575–599

Fujikawa K (2002) Load of carbon tax by region and income group. Input Output Anal Innov I-O Tech 10(4):35–41 (In Japanese)

Ministry of Finance Japan (2009) Outline of the FY2010 Tax Reform. http://www.kantei.go.jp/jp/kakugikettei/2009/1222zeiseitaikou.pdf

Ministry of Finance Japan (2010) Outline of the FY2011 Tax Reform. http://www.cao.go.jp/zeicho/etc/2010/__icsFiles/afieldfile/2010/12/20/221216taikou.pdf

Park S-J (2009) "Double dividend" of environmental tax reform. Koyosyobo, Tokyo (In Japanese)

Shimoda M, Watanabe T (2006) Re-examination of the scheduled carbon tax on the basis of IO analysis: a quantitative analysis on household burden by income class and by region. Bus Rev Aichi Gakuin Univ Bus Assoc Aichi Gakuin Univ 46(3):47–62 (In Japanese)

Sugino M, Arimura TH (2011) The effects of voluntary action plans on energy-saving investment: an empirical study of the Japanese manufacturing sector. Environ Econ Policy Stud 13(3):237–257

Sugino M, Arimura TH, Morita M (2012) The impact of a carbon tax on the industry and household: an input-output analysis. Environ Sci 25(2):126–133 (In Japanese)

Sugino M, Arimura TH, Morgenstern R (2013) The effects of alternative carbon mitigation policies on Japanese industries. Energy Policy 62:1254–1267

Takeda S, Kawasaki H, Ochiai K, Ban K (2010) An analysis of the medium-term target for CO_2 reduction by the JCER-CGE model. Rev Environ Econ Policy Stud 3(1):31–42 (In Japanese)

Takeda S, Arimura TH, Tamechika H, Fischer C, Fox AK (2014) Output-based allocation of emissions permits for mitigating the leakage and competitiveness issues for the Japanese economy. Environ Econ Policy Stud 16(1):89–110

Tokyo Tax Commission (2010) FY 2010 Tokyo tax commission interim report. http://www.metro.tokyo.jp/INET/KONDAN/2010/11/DATA/40kbt101.pdf (in Japanese)

Watanabe T (2009) Input–output analysis of regional economy. Seibundo, Kyoto (In Japanese)

Chapter 8
The Impact on the Japanese Economy of Reducing Greenhouse Gas Emissions: The Role of Economic Models

Abstract How much greenhouse gas (GHG) emissions should be reduced to mitigate their adverse effects on climate change is a global issue. As the first commitment period of the Kyoto Protocol (2008–2012) reached its end, a heated discussion continued over the post-Kyoto Protocol treaty, which urged the Japanese government to set Japan's medium-term target for limiting GHG emissions. Quantitative economic models were expected to play an essential role in this policy discussion. Chapter 8 describes the contributions that economic models can make to the policy debate over Japan's reduction target and offers some remarks on their use.

Keywords Climate Change Mitigation • Economic Model • Model Comparisons

8.1 Introduction

Before the opening of international negotiations for the post-Kyoto Protocol framework, Japan's medium-term target for reducing its greenhouse gas (GHG) emissions was discussed within the Japanese government. In 2008, the Aso Cabinet convened meetings of several experts from research institutions and professionals and established the Council on the Global Warming Issue within the prime minister's office. The council used economic models to conduct a feasibility study of Japan's medium-term target and an assessment of the likely effect of the target on the Japanese economy. Based on the Council's findings, the Aso Cabinet selected a 15 % reduction from the 2005 emissions level (or an 8 % reduction from the 1990 level) as Japan's medium-term goal.

After the formation of the Hatoyama administration, however, the Japanese government pledged to reduce Japan's GHG emissions by 25 % from the 1990 level on the condition that all major countries agree to ambitious targets. Just before the 2009 United Nations Climate Change Conference in Copenhagen (COP15), the Task Force of the Committee of Ministers on Global Warming met to assess the likely economic effects of the target. The effects of a 25 % reduction in GHG emissions on the economy, as predicted by economic models, were publicly

T.H. Arimura, K. Iwata, *An Evaluation of Japanese Environmental Regulations*,
DOI 10.1007/978-94-017-9947-8_8

presented and examined.[1] The issues identified were taken up for discussion by a subcommittee responsible for developing a road map for reaching the goal. At the meeting, there was a consensus that the economy would be able to grow with the 25 % reduction target.

The efforts made by both Prime Minister Aso's Council on the Global Warming Issue and Prime Minister Hatoyama's Task Force can be appreciated in the sense that they attempted to provide an objective assessment of the medium-term target for reducing GHG emissions by taking the likely economic consequences into account.

However, different economic models predicted the economic effects of GHG emission reductions differently. These differences seem to have caused some confusion for concerned parties, including the media.

This chapter discusses the use and limitations of economic models for quantitative analysis of the impact of a GHG emissions reduction target on the Japanese economy.

8.2 Economic Model Assessment of the Effects of GHG Emission Reductions on the Economy

Quantitative economic analysis can play an important role in policy development. In principle, a policy should not be implemented unless the benefits of the policy's implementation can be expected to exceed its costs. However, it is often difficult to quantify the expected benefits of a policy. An example of this challenge is the difficulty of quantifying the expected benefits of a reduction in GHG emissions.

8.2.1 The Benefits of GHG Emission Reductions

We wish to emphasize at the outset the significance of the benefits of GHG emission reductions. These benefits are associated with mitigating the adverse effects of climate change. For various reasons discussed below, the economic models that are used to assess Japan's policies regarding climate change do not provide estimates of the positive consequences of reduced GHG emissions.

First, considerable uncertainty exists about the effects of GHG reduction on the world's climate, and these effects are difficult to predict. Furthermore, how Japan's climate will change as the global climate changes is difficult to predict. For example, if we want to predict the effects of climate change on Japanese

[1] Relevant materials, including the minutes of this meeting, are available on the Cabinet website: http://www.kantei.go.jp/jp/singi/index.html.

agriculture, we need to predict how the temperature and rainfall in each region will change, which is challenging itself.

Second, even if we were able to predict climate change accurately, it would still be difficult to estimate the economic impact of climate change. Japanese agricultural products might change in response to climate change. Attempting to capture these reflected effects makes it difficult to assess the economic consequences of climate change. Attempting to predict the effects of climate change on our lifestyle is even more difficult. For example, suppose that people feel more uncomfortable in the hotter weather that results from climate change. This discomfort could be considered an economic cost. However, it is not straight forward to assign a precise monetary cost to increased human discomfort in hotter weather. The moderating effect of reduced GHG emissions on human discomfort in hotter weather could be considered an economic benefit, but its monetary value is also difficult to quantify.

Third, Japan's share of global GHG emissions may not be considered substantial. Therefore, the benefit of the Japanese emission reduction may not be sizable. In addition, Japan's reduction in emissions may be less beneficial for Japan than for other countries, which complicates the assessment of the benefits.

Thus, it seems difficult to assess the comprehensive benefits of reductions in GHG emissions. Therefore, our focus is on the effects on the economy of reducing GHG emissions to a target level. In other words, we focus on evaluating narrowly-defined benefits—namely, economic benefits or costs—and exclude environmental benefits from our analysis.

8.2.2 Economic Models for Evaluating the Economic Effects of GHG Emission Reduction Policies

Roughly speaking, two types of economic models exist for assessing the effects of GHG emissions reduction policies. The first type of model describes various energy use technologies in detail. The models developed by the Institute of Energy Economics, the AIM[2]/Enduse model developed by the National Institute for Environmental Studies and the DNE[3]21+ model developed by the Research Institution of Innovative Technology for the Earth all fall into this category. Models of this type focus primarily on the description of technology choices and their energy consumption. These models have the advantage of capturing in detail technology choices, their energy consumption and the resulting GHG emissions. A disadvantage of these models is that they are not suitable for assessing changes in economic activities as measured, for example, by GDP. Models of this type usually assume a certain level of economic activity and consider the choice of technology to be

[2] AIM is the abbreviation for Asian Pacific Integrated Model.

[3] DNE is the abbreviation for Dynamic New Earth.

consistent with the assumed economic state. Hence, the estimation of the production level of each industry comes from outside the model.

The second type of economic model seeks to comprehensively encompass economic activities. In particular, an applied general equilibrium model, also known as computable general equilibrium (CGE) model, describes the whole economic system and is able to capture fossil fuel consumption in detail. The CGE model developed by the Japan Center for Economic Research (JCER) (Takeda et al. 2010), the Keio Economic Observatory Model (KEO Model) and the AIM-CGE model all fall into this category. In an applied general equilibrium model, the economic activity level is endogenously determined in response to the implementation of carbon regulations. Thus, it is not necessary to make assumptions about each industry's activity level outside the model. CGE models provide a less arbitrary analysis of industrial activities than technology-oriented models, but they are not suitable for analyzing technological details. Thus one can conclude that applied general equilibrium models are more appropriate for evaluating economic impacts.

8.2.3 Economic Effect Assessment: Policy Comparisons

We can use various economic indicators to evaluate the economic effects of a mitigation policy. Typically, the effects on familiar economic indicators, such as GDP or household-related figures, are cited in policy discussions.

Here, we explain how we estimate the cost of introducing a GHG mitigation policy in terms of its negative effect on GDP. First, using a quantitative economic model, we exogenously assume an economic growth rate without carbon emissions regulation; based on this assumption, we forecast the GDP for 2020, which we call the baseline. Then, we predict how much the GDP will decrease as a result of the implementation of GHG emissions regulation. The difference between the GDP with and without the regulation is considered the cost imposed by the regulation.

Table 8.1 shows this cost estimation according to the JCER's CGE model. The model assumes a final target of a 25 % reduction in GHG emissions and estimates how much every 5 % reduction, from 10 to 25 %, would decrease GDP. In addition to a domestic reduction in emissions, the analysis assumes an overseas reduction through the use of carbon credit systems, such as the Clean Development Mechanism (CDM). Table 8.1 shows that a greater emissions reduction rate results in a

Table 8.1 Economic impact analysis based on JCER's CGE model

	Scenarios			
Domestic GHG emission reduction	10 %	15 %	20 %	25 %
Reduction in real GDP	0.8 %	1.3 %	2.1 %	3.1 %

From the Task Force Committee's Interim Report
The use of carbon credits is assumed in addition to a reduction in domestic GHG emissions

larger decrease in real GDP. The results suggest that a 25 % emissions reduction may decrease real GDP by 3.1 %.

The analysis results suggest that a combination of the lower domestic reduction target and the more use of carbon credits will alleviate the negative economic effects on Japan. As we have shown above, economic models work well for comparing several policy options (in this case, the use of carbon credits).

8.2.4 Factors Affecting the Magnitude of Economic Impacts

Caution is warranted here. The estimated economic impact may vary depending on the assumptions made. For example, if we assume a higher GDP growth rate for the base case, the decline in GDP will be larger, which implies a higher cost of emissions regulation. Differences in the assumptions made for the base case may result in large differences in the estimated costs.

Other external factors also affect the regulatory costs of reducing GHG emissions. For example, higher oil prices discourage oil consumption and thereby reduce GHG emissions. If oil prices rise, a model may predict that the amount of GHG emissions will decrease at a faster pace by 2020 without government intervention. Given initially higher oil prices, smaller efforts are required to achieve the reduction target, resulting in lower economic costs. By contrast, if oil prices decrease in the base-case scenario, the same model may predict that Japan will not have made substantial progress in reducing carbon gas emissions without regulation by 2020. As a result, Japan's economy will incur a larger additional cost to achieve the reduction target.

In addition, different ways of implementing GHG emissions regulation produce economic effects of different magnitudes. As we discuss later, decisions about how to spend carbon tax revenues affect the economy. Takeda et al. (2014) show that in attempts to reduce GHG emissions through emissions trading systems, adverse effects on energy-intensive industries can be significantly mitigated by adopting the output base allocation. As we have discussed, differences in policy assumptions result in different predictions of economic impact.

8.2.5 Comparisons of the Models

The predicted economic costs of reducing GHG emissions also depend on the assumptions made concerning the substitutability of various fuels and raw materials. For example, suppose that coal is currently used to produce certain products and that coal prices rise due to regulation of GHG emissions. In this case, the elasticity of substitution can serve as a measure of the ease or difficulty of switching from using coal to using liquid natural gas (LNG). Greater elasticity implies that the economy will transform itself at a faster pace into an economy that produces lower carbon gas emissions as a result of carbon taxes.

As we mentioned above, estimates of the economic costs associated with reduced emissions can vary substantially depending on the values of the parameters, such as the elasticity of substitution, in the economic model used. Although the values of a model's parameters are chosen to reflect the current state of the economy, the estimated economic costs are not necessarily identical because of differences in the structure of and assumptions associated with each model. For this reason, it would be inappropriate to compare the results obtained from several different models. We should not place undue emphasis on the estimated levels of economic variables such as GDP. A more effective use of these models is to compare the estimated economic consequences of various policy alternatives within the framework of each model.

When comparing the results of different models, we should keep in mind that their predicted values are only point estimates. For instance, at the Task Force, the JECR's CGE model, the KEO model and the AIM-CGE model predict 3.1 %, 5.6 %, and 3.2 % decreases in GDP, respectively, under regulation that reduces domestic carbon gas emissions by 25 %. However, we should not necessarily conclude based on these results that the models produce significantly different predictions. When we interpret these estimates, we should take into account their range or confidence interval because each model assumes different parameter values and has a different structure. Hence, by considering the confidence interval of the results, we may or may not conclude that the models yield different predictions of the impact of emissions regulation on the economy, even though the predictions may appear to be different.

8.2.6 The Importance of Effective Communication

Of course, we can conduct an analysis that considers the stochastic nature of parameter estimation. For this purpose, the use of an analytical method called *sensitivity analysis* is an option. This method enables us to see how changes in the parameter values affect the estimation results. A weakness of this method, however, is that its interpretation may not be straightforward for nonprofessionals because sensitivity analysis provides a range of estimated values rather than single point estimates. Hence, this method tends not to be adopted for political decision-making processes, and its use is largely limited to fields of academic research.

Researchers, members of the mass media and policymakers need to learn the communication skills needed to present scientific research results, such as those from sensitivity analyses, in an accurate but easily understandable way. How the results obtained from a model are explained to the public has a substantial influence on how the public interprets those results. A typical example in the context of the mid-term emission reduction target is the controversy over how much the emissions reduction target will cost the household sector.

The price for utilities, such as electric power and gas and for goods whose production involves energy consumption, is predicted to increase as a result of the GHG emission regulation. The mass media reported that the cost of living per household would increase by 360,000 yen if the government took measures to reduce GHG emissions by 25 %. This estimate was based on the results of an analysis conducted by the committee responsible for setting the medium-term emission reduction target. The magnitude of the predicted household cost-of-living increase raised public concern about the implementation of the medium-term target. The mass media's reporting of this predicted increase, however, was misleading in two respects.

First, the reported amount was the sum of a 220,000-yen decrease in disposable income (including consideration of the effect of energy price increases) and a 140,000-yen increase in energy bills (due to energy price increases). Thus, the effect of increases in energy prices was mistakenly double counted in media reports. This example illustrates the existence of communication problems between researchers and the mass media.

Second, the media reports gave the impression that each household would bear the cost-of-living increase under its current income circumstances, which might have made the public consider the cost of losing that amount immediately. Actually, this household cost-of-living increase was an estimate for the year 2020. Hence, if a household's income were to grow between 2010 and 2020 due to economic growth, the effect of the emissions reduction target would be a slower pace of household's income increase rather than the immediate decrease of the household's income.

As discussed above, miscommunication and misunderstandings may influence the nationwide discussion on setting the mid-term GHG emission target. Thus, great care must be taken in communication to develop a national consensus on emissions reduction.

8.3 Does Environmental Regulation Have Any Positive Effect on the Economy?

In discussions of mitigation policies, a view has emerged that environmental regulations will create new industries and promote economic growth. When Mr. Obama become the president of the United States and implemented the Green New Deal programs, this view attracted considerable attention and was hotly debated. With a few exceptions, environmental regulations increase costs for businesses, according to the traditional neoclassical economic theory. This is true because the model assumes that each firm rationally minimizes its costs. The typical assumption made in the neoclassical model is that each firm uses resources efficiently to produce its products, subject to several constraints. Environmental regulation is an additional constraint for firms that place restrictions on their choices of combinations of inputs, such as labor, energy and technology. Consequently, firms are forced to switch to more environmentally friendly, but usually more expensive, inputs. As a result, regulation that restricts GHG emissions will result in increased production costs for firms.

The same argument holds in a macroeconomic model such as a CGE model that aggregates all firms into one industry sector. The economic models used to assess the impact of regulatory measures that limit GHG emissions are based on the same assumption. A question that arises, then, is whether these regulations do any good.

8.3.1 Possible Positive Effects on the Economy: Double Dividends

If we do not take environmental problems into account, the social benefits of the production and consumption of goods and services are regarded as the sum of the benefits to the consumers and the benefits to the producers. As we explained in Chap. 1, "the competitive market" maximizes social welfare under certain conditions.

If there is some distortion in the market, the market mechanism cannot maximize the social welfare, even without negative externalities. In the real world, it is often the case that markets are distorted by several factors, including government intervention. For example, firms pay corporate taxes. While tax revenues are needed to support public functions, including central and local governments, corporate taxes prevent the maximization of social welfare in the economy in the sense that they reduce firms' incentives to make optimal investments. For this reason, corporate taxes are considered to have negative effects on market efficiency.

On the other hand, carbon taxes (i.e., taxes associated with GHG mitigation policies) can mitigate the adverse effect of corporate taxes and contribute to economic growth if the revenue is appropriately used. First, carbon taxes benefit the economy by limiting climate change. Second, if the resulting tax revenue is used to cut corporate taxes, firms may improve their investments. This is called the "double dividend" argument.

Allocating carbon tax revenue to decrease social insurance premiums rather than to reduce corporate taxes would have the same effect. For firms, expenses related to social security are costly and prevent them from creating more employment. Thus, if the revenues from carbon taxes were used to reduce the burden of social-security-related expenses, firms could hire more employees and the economy would grow. This type of tax reform was, in fact, implemented in Germany (Andersen and Ekins 2009).

However, quantitative analysis is needed to determine whether the net effect of the double dividend of a carbon tax is positive or negative. Corporate tax cuts and reductions in firms' expenses related to social security stimulate corporate activities, whereas carbon taxes restrict economic activity, at least in the short term, through reductions in GHG emissions. The former has a positive effect on the economy and the latter has a negative effect. Economic theories themselves are not sufficient to determine which effect is larger; thus, we need to estimate these positive and negative effects using quantitative economic models.

Estimation of the net amount of the double dividend has been attempted in Japan using CGE models. For instance, Tekeda (2007) study shows that if the revenue from carbon taxes were used to cut corporate taxes, the GDP would increase in the short and medium terms, with a GDP increase of 0.9 % expected 25 years after the introduction of the carbon tax.

Another study found that the carbon tax revenue allocated to reducing taxes on labor, which firms pay in the form of a social security premium on wages, would increase the GDP (Takeda et al. 2014). Furthermore, according to Kawase et al. (2003), GDP would increase if carbon tax revenue were used to cut the consumption tax.

As discussed above, the results of model-based analyses suggest the possibility that environmental regulation will promote economic growth due to the double dividend effect of carbon taxes. These results also suggest that the expected economic improvement is, at most, 1 % of GDP.

8.3.2 Possible Positive Effects of Regulation: The Porter Hypothesis

The Porter hypothesis proposes another possibility for how environmental regulations may encourage economic growth. Porter argues that the enforcement of strict environmental regulations motivates firms to invest in research and development (R&D) to find and develop new technologies to cope with the new regulatory environment—new technologies that may never have been previously imagined. Innovations that result from firms' R&D efforts improve firms' international competitiveness (Porter 1991).

Individual cases that support Porter's hypothesis have been reported. However, the hypothesis must be tested on the economy as a whole before its effects can be incorporated into economic models. Empirical studies that have examined the validity of Porter's hypothesis in Japan have confirmed that stricter environmental regulations tended to increase firms' spending on R&D in the 1970s and 1980s (Hamamoto 1997). However, other research reports that in more recent years, this relationship has not necessarily held (Arimua and Sugino 2007). In addition, some recent studies based on firm-level data have suggested that environmental regulations may increase a firm's environment-related R&D spending at the expense of R&D spending in other areas (Arimura and Sugino 2008). In other words, without the implementation of environmental regulations, funds spent on environment-related R&D could have gone to R&D in other areas that might have yielded greater benefits for the firm. To assess the effects of environmental regulations, we need to assess both the benefits due to environment-related innovations and the loss in benefits from other innovations that might otherwise have been developed.

As we have observed, no consensus exists on how Porter's hypothesis can be quantified in Japanese economy, and at this stage, it seems difficult to integrate this hypothesis into economic models used to analyze the effects of GHG emission regulations.

8.3.3 Japanese Product Sales in Overseas Markets

Even if Porter's hypothesis does not hold, environmental regulations can work in favor of the Japanese economy. For example, Japan may be able to export its environmentally friendly products to foreign countries with regulations similar to Japan's if the Japanese products meet these countries' environmental standards.

Many models used to evaluate the economic impacts of GHG emission regulations have been developed only to analyze the effects on the domestic economy. Thus, these models are not suitable for examining spillover effects in relation to overseas markets. Furthermore, to measure the disseminating effects of a specific energy-saving product in the global market, we need to construct a comprehensive model that encompasses detailed subsector models for each industry's goods and services.

Mitigation policies affect the economy over the medium and long term. Therefore, we need to make predictions to assess the effects of such policies. However, it is difficult to develop a long-term prediction model that takes into account the international effects of environmental regulations. In reality, we must make a choice between constructing a model that focuses on the domestic economy for the purpose of making predictions and constructing a model that focuses on the economy at one point in time for the purpose of considering international responses.

8.3.4 Green Consumption and Green Industry

A conceivable exceptional case is one in which consumer preferences become greener due to environmental regulations, leading to greener consumption habits. If consumers prefer low-carbon-emission products, environmental regulations increase social economic welfare, even in a narrow sense. However, at this stage, we do not have enough evidence to support the validity of this argument. To determine whether and how this concept should be incorporated into economic models, we believe that further empirical studies should be conducted.

Some researchers hold a view that environmental regulations should be strengthened because tighter regulations create jobs in the "green industry." However, we need to examine this assertion carefully. First, if job creation in the green industry depends on subsidies, as is currently the case for solar power generation, we should scrutinize where the money for the subsidies is justified or not; it may be true that many more jobs could have been created if the money had been spent on other industries. Second, the rise of the green industry may result in the fall of some other existing industries. Thus, we need to assess the effects of environmental regulations on the economy as a whole rather than on individual industries alone.

8.4 Limitations of Economic Models

It is necessary to understand the limitations of economic model analysis. We restrict our attention here to the following points.

8.4.1 Assessment of Effects on the Household and Transportation Sectors

The study of the effects of emissions regulations using traditional economic models, which are developed to describe the workings of an economy as a whole, has been focused on the analysis of the effects of these regulations on industrial activities. These models do not take into detailed consideration energy consumption in the household and transportation sectors. For example, the structure of cities and the availability of public transportation affect the amount of GHG emissions by the transportation sector. The amount of GHG emissions is also a variable in the network structure of a complex transportation system and a variety of transport policies, such as the mitigation of traffic congestion on roads. However, many models do not address these issues because they do not take spatial factors into account. With respect to the household sector, the structure of energy consumption depends on the type of housing: detached houses versus multi-family units. Traditional models do not capture the differences in energy consumption patterns for different types of housing structures.

Looking forward, issues concerning how environmental regulation affects the household sector and the transportation sector will need to be addressed in refinements to the economic models discussed here.

8.4.2 The Difficulty of Evaluating an Energy/Carbon Intensity Target

With respect to the choice of policy target, a specific level of energy or carbon intensity is often set as a target for reducing GHG emissions rather than a total energy consumption or carbon emission reduction such as a 25 % cut. Energy or carbon intensity is expressed as energy consumption or carbon emission per output.

Traditional economic models are not built on the assumption that they will address a specific energy consumption target rather than intensity targets.[4]

In addition, traditional models have not been sufficiently refined to estimate and define the costs of improvement in energy intensity. Consequently, these models are not capable of analyzing the effectiveness of measures such as the voluntary action plan that Japan Business Federation (Keidanren) has been implementing. If economic models could address energy intensity improvement, they would be able to contribute more to discussions about GHG emission regulations.

8.4.3 *Assessment of Green Innovation*

Economic models are often not good at assessing drastic changes such as innovative technologies and new products that did not exist in the past. Consequently, they may not be suitable for the evaluation of green consumption and green investment, at least at this stage. However, some study argues that Porter's hypothesis can be explained using an economic model that assumes neoclassical rationality while remaining consistent with traditional models. For a case in which there is uncertainty concerning the outcomes of R&D, Porter's hypothesis can hold even in a neoclassical model (Popp 2005).

For this argument to be true, however, groundbreaking innovation is required; a gradual increase in R&D is not enough. Thus, in reality, it would be implausible for Porter's hypothesis to hold without the implementation of exceptionally strict environmental regulations.

In essence, with a few exceptions, we have not yet experienced strict environmental regulations that can lead to drastic innovation. Thus, we have not yet obtained conclusive results concerning the validity of Porter's hypothesis. This amounts to a "chicken or egg" argument. Thus, it remains difficult for currently available economic models to consider innovative technologies based on Porter's hypothesis.

8.5 Conclusions

This chapter explained the role, the state, and the limitations of economic model analysis as it applies to environmental regulation. It should also be noted that the development of an economic model requires considerable time and human resources. Building an economic model and analyzing policy are time-consuming and human-resource intensive activities. In contrast, policy decisions are often

[4] One exception is Sugino and Arimura (2011). They examined the impacts of intensity target on investment in Japanese manufacturing sectors.

made rapidly. There seems to be a large gap between researchers' and policymakers' sense of time. Analysts should endeavor to close the gap between the two, while policymakers are expected to understand that model analysis is time-consuming.

References

Andersen MS, Ekins P (2009) Carbon energy taxation – lessons from Europe. Oxford University Press, Oxford

Arimura TH, Sugino M (2007) Does stringent environmental regulation stimulate environment related technological innovation? Sophia Econ Rev 52:1–14

Arimura TH, Sugino M (2008) The impact of environmental regulation on technological innovation: an assessment of the porter hypothesis. J Sci Policy Res Manage 23(3):201–211 (In Japanese)

Hamamoto M (1997) Reconsideration and empirical analysis of the porter hypothesis. Econ Rev 160(5–6):102–120

Kawase A, Kitaura Y, Hashimoto K (2003) Environmental tax and the double dividend: a computational general equilibrium analysis. Tokyo Pub Choice Stud 41:5–23 (in Japanese)

Popp D (2005) Uncertain R&D and the porter hypothesis. BE J Econ Anal Policy 4(1):1–14

Porter ME (1991) American's green strategy. Sci Am 264(4):96

Sugino M, Arimura TH (2011) The effects of voluntary action plans on energy-saving investment: an empirical study of the Japanese manufacturing sector. Environ Econ Policy Stud 13(3):237–257

Takeda S (2007) The double dividend from carbon regulations in Japan. J Jpn Int Econ 21(3):336–364

Takeda S, Kawasaki H, Ochiai K, Ban K (2010) An analysis of the medium-term target for CO_2 reduction by the JCER-CGE model. Rev Environ Econ Policy Stud 3(1):31–42 (In Japanese)

Takeda S, Arimura TH, Tamechika T, Fischer C, Fox A (2014) Output-based allocation of emissions permits for mitigating the leakage and competitiveness issues for the Japanese economy. Environ Econ Policy Stud 16(1):89–110

Index

A
Air pollution, 1, 2, 7, 13, 15, 20–23, 38, 51–53, 60, 64, 68, 95, 96, 103, 104, 116
Ancillary benefits and costs, 7–8
Average life expectancy, 26, 27, 33, 47–48
Aviation fuel tax, 147, 149, 150

B
Box–Cox transformation, 77, 80

C
Carbon tax, 7, 16, 140, 146–159, 161–167, 179, 180, 182, 183
Clean development mechanism, 139
Compliance cost, 20, 21, 24, 26, 29–30, 40, 46, 47, 57
Comprehensive energy management, 126–128, 140
Computable general equilibrium model, 178
Congestion, 16, 76, 88, 92, 95, 99–103, 106–108, 114, 116, 118, 119, 185
Corporate average fuel economy, 52
Cost benefit analysis, 11–13, 15, 20, 21, 24, 38, 39, 51–69
Cost effectiveness analysis, 9, 11, 12

D
Diesel particulate filter, 15, 55
Double dividend, 182, 183

E
Economic incentives, 3, 4, 6, 13, 45
Economic model assessment, 176–181
Efficiency, 1, 6, 7, 12, 32, 39–46, 52, 71, 101, 116, 124, 126, 128, 129, 131–135, 138–140, 146, 151, 166, 182
Electricity tax, 158, 159
Electric toll collection, 87
Emission
  intensity, 55, 56, 60–62, 64, 67, 68
  standard, 9, 21, 36, 51, 71
Energy
  intensity, 16, 124, 126, 128, 130, 131, 134–140, 186
  management, 125–129, 132–137, 139, 140
  management facilities, 125, 129, 132
Energy Conservation Act (Act, rational use of energy), 124–129, 131, 132, 139, 140
Energy-related tax, 146–151
Environmental tax, 4–8, 45, 46, 139, 146, 147, 151, 155
Expressway toll system, 97
Externality, 1, 3, 4, 15, 20, 41, 63, 64, 147, 150, 151

F
Fleet scrappage program, 72

G
Gasoline tax, 52, 115, 147–150
General equilibrium approach, 89

T.H. Arimura, K. Iwata, *An Evaluation of Japanese Environmental Regulations*,
DOI 10.1007/978-94-017-9947-8

Generalized cost, 93–96, 104–109, 112, 113
Generalized equilibrium demand, 91–92
Government Policy Evaluation Act (GPEA), 13, 14
Grace period, 54, 55, 68
Green consumption, 184, 186
Green industry, 184–185
Green innovation, 186

H
Heat consumption, 130, 131, 134–136, 138, 141
Hedonic approach, 76
Household expenditure, 162–165

I
Input-output analysis, 151–153, 161
Iso-net benefit, 66, 67

K
Kyoto protocol, 16, 123, 124, 127, 128, 139, 145, 175

L
Light oil delivery tax, 147, 149, 150
Liquefied Petroleum Gas Tax, 147, 149

M
Marginal abatement cost, 6, 39–41, 139, 140
Marginal external cost, 3, 37
Market failure, 1

N
Natural replacement, 31, 34, 42, 48
Nitrogen oxide, 2, 20–22, 51, 72, 73
NOx-PM act, 6, 7, 15, 22, 25, 46, 52, 54, 71–84, 89

O
On-site inspection, 127, 128, 129
Operational regulation, 52–64, 66
Opportunity cost, 25, 27, 29, 46, 58, 93, 112
Optimization problem, 42, 45
Outflow, 15, 72, 73, 83

P
Particulate matter, 2, 20, 22, 51–69, 72
PDCA cycle, 8, 9
Petroleum and Coal Tax, 147, 149
Policy comparison, 178
Pollution haven hypothesis, 72, 84
Porter hypothesis, 183
Prescriptive regulations, 3, 4, 6, 13, 52
Price increase, 153, 156–162, 167–170, 181
Problem of equality, 151
Promotion of power resource development tax, 147
Public announcement, 127, 128

Q
Quantile regression, 138, 140–143
Quantitative evaluation, 6, 9, 10

R
Regulatory impact analysis, 8
Robustness check, 53, 65–68

S
Secondary market, 15, 78, 84
Sensitivity analysis, 13, 37, 38, 65, 66, 69, 180
Small and medium-sized facilities, 139
Social cost, 3, 6, 11, 12, 14, 55, 57, 89, 91–96, 107–111, 113, 116
Social welfare, 2–3, 5–8, 10, 12, 16, 53, 67, 68, 72, 89, 91, 110, 116, 150, 182

T
Tax exemption, 151, 155–157, 159, 161, 162, 166
Terminal year, 20, 22, 25–29, 31, 39, 41–46, 49, 52, 53, 55, 63, 64, 72
Tokaido Shinkansen, 89, 92, 111–116
Tokyo Metropolitan government, 148
Tomei expressway, 88, 89, 91–93, 95, 97–100, 104–112, 114–116, 118–121
Traffic congestion, 16, 88, 92, 95, 99–101, 103, 106, 108, 114, 116, 118, 119, 185

U
Uncertainty, 37, 65, 176

V
Vehicle type regulation, 19–49, 52–56

W
Welfare analysis, 87–121

Y
1000-Yen Expressway Discount Program, 88

MIX
Papier aus verantwortungsvollen Quellen
Paper from responsible sources
FSC® C105338

If you have any concerns about our products,
you can contact us on
ProductSafety@springernature.com

In case Publisher is established outside the EU,
the EU authorized representative is:
Springer Nature Customer Service Center GmbH
Europaplatz 3, 69115 Heidelberg, Germany

Printed by Libri Plureos GmbH
in Hamburg, Germany